Indian Dairying

NIPA® GENX ELECTRONIC RESOURCES & SOLUTIONS P. LTD.
New Delhi-110 034

Indian Dairying
Challenges and Opportunities

Vijay Kumar
Assistant Professor
Department of Animal Genetics & Breeding
College of Veterinary Science & Animal Husbandry
U.P. Pandit Deen Dayal Upadhyaya Pashu Chikitsa Vigyan Vishwavidyalaya
Evam Go Anusandhan Sansthan (DUVASU)
Mathura, Uttar Pradesh

NIPA® GENX ELECTRONIC RESOURCES & SOLUTIONS P. LTD.
New Delhi-110 034

NIPA® GENX ELECTRONIC RESOURCES & SOLUTIONS P. LTD.

101,103, Vikas Surya Plaza, CU Block
L.S.C.Market, Pitam Pura, New Delhi-110 034
Ph. +91 11 27341616, 27341717, 27341718
E-mail: newindiapublishingagency@gmail.com
www: www.nipabooks.com

For customer assistance, please contact
Phone: + 91-11-27 34 17 17
Fax: + 91-11-27 34 16 16
E-Mail: feedbacks@nipabooks.com

ISBN: 978-81-19002-46-7

Composed and Designed by NIPA®.

उ.प्र.पं. दीनदयाल उपाध्याय पशु चिकित्सा विज्ञान विश्वविद्यालय
एवं गो अनुसंधान संस्थान, मथुरा-281001 (उ.प्र)

U.P. Pandit Deen Dayal Upadhyaya Pashu Chikitsa Vigyan Vishwavidyalaya Evam Go Anusandhan Sansthan (DUVASU) Mathura-281001 (U.P.)

प्रो.(डॉ.) ऐ.के. श्रीवास्तव
कुलपति
Prof. (Dr.) A.K. Srivastava
Vice-Chancellor

Foreword

I am immensely delighted to write a foreword for the book entitled, "Indian Dairying: Challenges and Opportunities" written by Dr. Vijay Kumar of Uttar Pradesh Pandit Deen Dayal Upadhyaya Pashu Chikitsa Vigyan Vishwavidyalaya Evam Go-Anusandhan Sansthan, Mathura. India's journey from a milk deficit country to milk surplus nation has been historic. Indian dairy industry is symbolic to the upward trajectory of the country's economic power and influence. Consumers with growing socio-economic status and more income are looking for healthy food alternatives. As a consequence, milk and milk products are growing in popularity. Although, India has emerged as the highest milk producer in the world and dairy industry has played a very significant role in economic, social and nutritional well-being of rural masses. There are many challenges for Indian dairy sector. In spite of widespread popularity of Indian dairy products, India is a very minor player in the world dairy food market. There are several reasons responsible for this situation such as slow breed improvement, shortage of feed and fodder, emerging new disease, poor penetration of technologies at farmers' door, inadequacy of appropriate technologies for commercial production, lack of quality assurance system and more important nonexistence of industry-institution linkage.

The objective of the book is to enlighten stake holders on different important issues on dairy sector of India. I am confident that this book will go a long way in helping the students, dairy farmers, entrepreneurs, academicians, researchers and other stake holders of the dairy sector. I wish the author all the best.

AKSrivastava.

A K Srivastava

Tel. : 0565-2470199; Fax: 0565-2470819; Mob. 9466592661; Email: duvasuvc@gmail.com
https://www.upvetuniv.edu.in

Preface

India has an enormous wealth of cattle, which is crucial to the development of the socioeconomic status of millions of rural households as well as the country's economy. In addition to giving millions of people access to low-cost nutritious food, this industry plays a significant role in boosting family incomes and creating gainful employment in rural areas, notably among landless laborers, small and marginal farmers, and the women. The introduction of dairying has been beneficial for dairy producers and particularly significant for those societal groups that have historically been weak. It has given a year-round source of income where they previously could only rely on revenues from seasonal crops or irregular labor. Dairying has developed into a full-fledged industry in the nation, positively enhancing the lives of those involved in the industry, whether directly or indirectly, and bringing about significant socioeconomic improvements. With expanding population, growing urbanization and increasing purchasing power, the demand for milk and milk products is expected to go up. We need to catch up with the daunting challenge to produce more milk and convert it efficiently into value added products. Global trend in animal production also indicates a rapid and massive increase in consumption of dairy products. To meet the domestic demand of milk and to generate sufficient exportable surplus, the growth rate in our milk production needs to be accelerated.

Indian dairy research priorities need to be focused on development of technologies in the area of dairy science viz. animal production, milk processing and extension and marketing. There is a need to increase the number of animals with superior germplasm, at the same time the number of non-producing and less producing animals should be reduced in order to minimize the competition on limited feed and fodder. Further the major challenges which require urgent attention are, low productivity, breed improvement, shortage of feed and fodder, emerging new disease, poor penetration of technologies at farmers door etc. To unpack these issues being faced by Indian dairy sector, it is pertinent to dive deep and list out the factors that have been hampering the productivity, processing and marketing of milk and dairy products.

I am highly thankful to Prof. A K Srivastava, Vice Chancellor, DUVASU, Mathura for appreciating my effort. I am also thankful to the students and teachers who read my two books on Animal Genetics & Breeding and encouraged me to write a book on Indian Dairying. I am also highly thankful to my family members – my wife Kajal, my son, my mother and my elder brothers & sister.

I hope this book will be of great use and help to the students, dairy farmers & entrepreneurs, dairy industry personnel, academicians, researchers and stake holders in Indian dairy sector.

Vijay Kumar

Contents

1

Livestock Production Statistics of India

The world's largest population of milch cattle lives in India. In 2020, there were almost 906 million tones of milk produced worldwide, with India being the top producer with a 22 percent share. Annual Growth Rate of milk production in 2019-20 in India was 5.69 per cent. Highest growth rate in milk production was found to be 14.31 per cent in Karnataka. In 2021, the Indian dairy market had a value of INR 13,174 billion. By 2027, the market is expected to reach INR 30,840 billion. Since 1998, with a steady increase in the supply of milk and milk derivatives, India has been the world's largest producer and consumer of dairy products. The rural Indian economy is heavily reliant on the dairy sector as a source of jobs and revenue. India also has the largest population of bovine in the entire world. Nevertheless, compared to the other major dairy producers, the milk production per animal is much lower. The majority of the dairy products supplied in India are fluid milk, and almost all of them are consumed domestically. Because of this, there is enormous opportunity for value addition and overall growth in the Indian dairy sector.

Table 1: Standings of India in the World Ranking

Rank	Sector
1st	Total Livestock Population, Milk Production, Buffalo Population, Goat Milk Production, Total Bovine Population
2nd	Cattle Population, Goat Population

Table 2: Average Yield per In-Milk Animal in 2019-20

Type of animal	Average yield
Exotic cow	11.88 kg/day
Cross-bred cow	8.09 kg/day
Indigenous cow	3.90 kg/day
Non-descript cow	2.57 kg/day

Share of States in Production of Livestock Products in 2019-20

- Largest Share in Total Milk Production in India: Uttar Pradesh (16.06%)
- Second Largest Share in Total Milk Production in India: Rajasthan (12.89%)

Key Highlights of 20th Livestock Census

- Total livestock population: 536.76 million (showed an increase of 4.6% over Livestock Census-2012)
- Total poultry population: 851.81 million (showed an increase of 16.8% over Livestock Census-2012)
- Milch animals: 125.75 million
- Total bovine population: 303.76 million
- Total cattle population: 193.46 million
- Total indigenous/non-descript cattle: 142.11 million
- Total exotic/crossbred cattle: 51.36 million
- Total buffalo population: 109.85 million
- Total goat population: 148.88 million

Table 3: Percentage Distribution of Total Livestock as per 20th Livestock Census

Species	Percentage
Cattle	36.04 per cent
Buffalo	20.47 per cent
Sheep	13.83 per cent
Goat	27.74 per cent
Pigs	1.69 per cent
Others	0.23 per cent

Table 4: Species Wise Milk Contribution to Total Milk Production in India

Species	Percentage of total milk
Buffalo indigenous	34.51 per cent
Buffalo non-descript	13.83 per cent
Cow non-descript	10.42 per cent
Cow indigenous	9.63 per cent
Cow cross-bred	27.68 per cent
Cow exotic	0.98 per cent
Goat	2.95 per cent

Table 5: Percentage of the World Livestock Present in India

Species	Percentage present in India
Total Livestock	11.54 per cent
Buffalo	56.7 per cent
Small ruminants	20.4 per cent
Cattle	12.5 per cent

Indian Dairy Production Statistics

Worldwide, dairy consumption accounts for a sizable portion of the human diet. Calcium, minerals, and proteins are just a few of the nutrients it provides. However, the usage of dairy in India is much more extensive than dietary and nutritional. It has a significant role in Hindu mythology. Cows are revered by Hindus as sacred representations of the goddess Kamdhenu. The Indian subcontinent uses milk and other dairy products since ancient time, according to the Rigveda. India is one of the world's top producers and consumers of milk in part because of this. Value of 66.55 per cent of total output from Livestock Rearing Sector is coming from Milk and Milk Products in India. In last 3 decades, world milk production has increased from 530 MT in (1988-89) to 937 MT in 2021-22. India is world largest milk producer, with 22% of global production, followed by the USA, China, Pakistan and Brazil. Since 1970s, most of expansion in milk production has been in south Asia. In 2020-21, milk production in Asia rose to 379 MT, mainly from India, China and Pakistan.

Table 6: Milk production and per capita availability of milk in India

Year	Production (Million Tones)	Per Capita Availability (gms/day)
2010-11	121.8	281
2011-12	127.9	290
2012-13	132.4	299
2013-14	137.7	307
2014-15	146.3	322
2015-16	155.5	337
2016-17	165.4	355
2017-18	176.3	375
2018-19	187.7	394
2019-20	198.4	406
2020-21	210	428

Source: Basic Animal Husbandry Statistics, DAHD&F, GoI

Important facts of Indian dairy sector

- The number of crossbred and indigenous cattle female increased by 10 and 6 per cent respectively from 2012 to 2019.
- The recommended per capita milk in India is 280 gm/day. Per capita milk availability in India is 428 gm/day. Global per capita milk availability is 229 gm/day.
- In 2021, dairy market in India reached to Rs 13,174 billion. By 2027, it is expected to reach to Rs 30,840 billion (15% annual growth).

- By 2030 the demand of dairy products is expected to grow at rate of 9-12%.
- Technology driven production, processing and value addition is the important option for the future.
- Dairy sector is growing faster than crop, horticulture and any other agricultural sub sector.
- By 2030, dairy sector is predicted to produce more than half of total agricultural output.
- In 2019-20, the growth in world milk production was 1.3%, however in India it was 6.5%.
- Milk is India's largest agricultural commodity. The value of milk is more than the total value of all grains and pulses put together (2019-20). The value of milk was more than Rs 8 lac crores as compared to Rs 4.5 lac crores of all grains and pulses.
- Today, white revolution's contribution surpasses green revolution. Every 4th rupee generated in farm sector comes from milk and 3rd rupee from livestock sector.

2

Indian Dairy Sector

India has an enormous wealth of cattle, which is crucial to the development of the socioeconomic status of millions of rural households as well as the country's economy. In addition to giving millions of people access to low-cost nutritious food, this industry plays a significant role in boosting family incomes and creating gainful employment in rural areas, notably among landless laborers, small and marginal farmers, and women. The best defense against the whims of nature, such as famine, drought, and other natural disasters, is livestock. With more than 56 percent of the world's buffaloes and more than 12 % of its cattle, India has one of the largest herds of both cattle and buffaloes. The world's largest producer of milk is India. The government has taken a number of steps to boost livestock productivity, which has led to a rise in milk production. From the 9th Plan onward, the Indian dairy industry experienced significant growth. This has not only elevated the nation to the top of the world, but also signifies a consistent improvement in the population's access to milk and milk products. Millions of rural families now rely heavily on dairying as a secondary source of income, and it now plays a key role in creating job and income-generating opportunities.

Significance of Indian Dairy Industry

The introduction of dairying has been beneficial for dairy producers and particularly significant for those societal groups that have historically been weak, such as small landholders, laborers without access to land, and women. It has given the community a year-round source of income where they previously could only rely on revenues from seasonal crops or irregular labor. It is estimated that dairying currently accounts for 60–65 percent of this group's revenue. Dairying has developed into a fully fledged industry in the nation throughout time, positively enhancing the lives of those involved in the industry, whether directly or indirectly, and bringing about significant socioeconomic improvements. Numerous studies have demonstrated that dairying significantly increases the income of the rural household in milk sheds. Since dairying is a secondary occupation for the majority of rural residents, the money they earn not only allows them to escape household poverty but also improves their social and economic standing. Dairy farming has also given rural residents, particularly widows and those without land, a place to live and a source of self-employment at no additional expense. The dairy industry is primarily run and owned by the farmers themselves. They can therefore utilize support services including veterinary help, artificial intelligence, and extension services. Additionally, this has led to a decrease in production costs. The financial situation of rural farmers improved as a result of this indirect benefit. The middlemen have been eliminated, which has curbed abuse of farmers, and a regular, lucrative market has been established. In turn, this has helped humanize their economy by reducing the distress sales of milk and milch animals. Members of livestock owners rely heavily on livestock products like milk as a source of animal nutrition. Milk is often available in quantities of 428 g per person per day. In terms of the owners' standing in society, the animals provide social security. Families, especially those without access to land, who possess animals, are in a better position than those who do not. In many regions of the nation, giving animals as wedding gifts is a highly widespread practice. Animal rearing is an element of Indian culture. Animals are employed for a variety of social and religious purposes. Cows are used in house-warming rituals. During numerous religious ceremonies, bulls and cows are worshipped. Many pet owners grow attached to their creatures. Gender equity is promoted by animal husbandry. Women provide more than three-fourths of the labor needed to produce animals. In Punjab and Haryana, where dairying is a key activity and animals are fed in stalls, the percentage of women employed in the livestock sector is approximately 90%.

Dairy Development in India

India was one of the major importers of dairy products after gaining independence in the 1950s, accounting for more than 40% of the global throughput of the dairy industry in terms of milk solids. In 1963–1964, the maximum amount of marketable milk powder imports was 53 thousand tones. This alarmed policymakers, who then decided to increase milk production. The National Dairy Development Board (NDDB), which was established in the middle of the 1960s to oversee the nation's dairy development, was a crucial step forward that had a broad impact. One of the largest and most successful dairy development initiatives in the world, Operation Flood (OF), was established in 1970 with the primary goal of establishing farmer cooperatives in rural areas and connecting them to urban consumers. The dairy industry in India has changed as a result of Operation Flood, and a reliable system for milk procurement, processing, and distribution by the cooperative sector has been established. The Technology Mission on Dairy Development (TMDD) initiative was started by the Indian government in 1989 to continue and supplement Operation Flood's efforts as well as to expand dairy farming as a source of income and employment in rural areas.

Agriculture's share of the GDP decreased from 34% in 1981–1982 to 15% in 2010–2011. Even while it didn't fall as dramatically as the size of the agricultural sector, livestock's contribution to GDP did. Up until 2000–2001, it fluctuated between 5–6% before gradually dropping to 3.9% in 2010–2011. Even Nevertheless, the livestock sector's contribution to the agricultural GDP increased steadily, rising from 15% in 1981–1982 to 26% in 2010–2011. In the 1980s, the livestock sector increased at a 5.3% annual pace, about twice as fast as the crop sector. Growth in the livestock industry slowed in subsequent decades and reached 3.6% in the 2000s. Despite this slowdown, the livestock industry's growth remained almost 1.5 times stronger than that of the crop sector, which supported total agricultural growth. However, the contribution of livestock to agricultural income varies significantly by location. Milk is the primary product of the livestock industry, accounting for 66.7% of the value of all animal output. The biggest agricultural product in India is milk. It directly employs 80 million dairy farmers and has a vital contribution to the national economy. The dairy sector is expected to rise by 9–11% in coming years as a result of a recovery in economic activity, rising per capita consumption of milk and milk products, shifting dietary choices, and increasing urbanization in India.

Over the previous five years ending in 2020, the cattle sector expanded at a Compound Annual Growth Rate (CAGR) of 8.15%. Over 50% of the dairy industry's revenue comes from the liquid milk segment, where growth is expected to be consistent (6-7%). Compared to the unorganized component, which makes up 26–30% of the business, the organized dairy segment has had faster growth. The dairy business in India is well-positioned for impressive growth in the ensuing decades. The factors that contributed to the increase in milk consumption—population expansion, urbanization, wealth growth, and dietary changes—are predicted to continue well into the next millennium, ushering in a true Livestock Revolution. The rise of milk producing activities in peri-urban areas is a result of the rapidly rising demand for dairy products in developing nations. Industrial dairy farms are rapidly meeting the demand for milk and dairy products in urban and peri-urban regions. In India, traditionally, crop productivity and milk production have been closely linked. Large commercial dairy farms operating in metropolitan areas could lead to tough competition to rural dairy. The rural areas do not receive the growth impetus from the high demand for animal products, which would promote rural development, more equitable income distribution, and the reduction of poverty. It is challenging for small producers to compete with huge commercial entities. The expansion of commercial dairy farms has a negative impact on the additional income and employment opportunities provided by milk production in rural communities. Crucial policy advancement was the Operation Flood initiative, which established a market link between urban milk consumers and rural milk producers via a web of cooperatives.

High non descript population

For the first time in a long time, India's cows have overtaken buffaloes, driving up the country's overall milk production to record highs. According to the most recent data published by the Department of Animal Husbandry and Dairying (DAHD), India produced 209 million tones of milk (provisional data), an all-time high, more than half of which came from cows. Up until this point, buffaloes had been the main producers. For instance, in 2016–17, they contributed almost 49.2% of India's 165.4 million tones of milk production, compared to cows' 47.3% and goats' 3.5%. But according to data, cows produced 51% of the country's total milk production in 2020–21, buffaloes produced 45% of it, and goats produced 3%. This represents a 29% increase in production over the previous five years. However, the proportion of indigenous and non descript cows has stayed relatively constant (20%), meaning that, relative to their population, exotic and crossbred cows account for a bigger share of milk production. Exotic and crossbred cows produce 28% of the country's milk. On the other hand, indigenous and non descript desi cows account for only 20.5% of the milk yield while making about 38% of the entire milk population. By providing farmers with a subsidy for in vitro fertilization (IVF) and sex-sorted semen that only produces female bovine animals, the government of India launched the Rashtriya Gokul Mission in 2014 to aid in increasing the productivity of indigenous cattle. This reduces the risk associated with investments in cattle. According to the DAHD's annual report for 2020–21, Sex-sorted semen production technology has been established in India for indigenous breeds of cattle such as Red Sindhi, Tharparkar, Sahiwal, Gir, etc.

According to DAHD statistics, a crossbred cow often produces 7-8 kg of milk per day, more than twice as much as indigenous and non descript desi cows, who only produce 2.4–3 kg. Due to the fact that these exotic cows require special care and are more susceptible to disease and the impacts of climate change, the contribution of exotic cows is just a few percent of total cow production. However, as the name would imply, these cows produce the most milk, averaging more than 9 kg per day. Therefore, crossbred cows are the best of both worlds since they produce more than desi cows while also being simpler to care for than exotic cows. Better economics and the emphasis on promoting this technology have both contributed to the rising share of the crossbred population and their milk production. Crossbred has the highest returns per rupee spent in terms of milk output, particularly during the first five lactations. Another indicator of these positive improvements in the dairy industry is the increase in the population of cattle. The total number of cattle (cows, buffaloes, yaks, and mithuns) increased by only 1% between 2012 and 2019, according to the 19th and 20th rounds of the livestock census, while the population of exotic and crossbred cows increased by 26% and that of indigenous and non descript cows by 10%.

Table 7: Animal-wise yield (kg per day) Provisional estimates for 2020-21, DAHD

Exotic cows	9.9 kg
Crossbred cows	7.2 kg
Indigenous buffaloes	6.4 kg
Non descript buffaloes	4.1 kg
Indigenous cows	3.3 kg
Non descript cows	2.7 kg
Goat	0.5 kg

While crossbred and exotic cows can be credited for pushing India's milk production levels to an all-time high, buffaloes have been the backbone of India's milk supply. According to the DAHD statistics, both indigenous and nondescript buffaloes yield about 4-6 kg of milk per day, which is higher than what indigenous and nondescript cows offer — about 2.4 to 3 kg per day. Farmers are increasingly choosing crossbred cows because of their higher production. Consumers hardly ever prefer to drink milk that has been cross-bred. Milk from crossbred animals is typically sold for processing to halwas and other eateries. Most liquid milk consumed comes from local cows and buffalo. The quality of buffalo milk is significantly superior in terms of fat and SNF (solids not fat), which supports the idea that buffalo's milk is more well-liked. While considering the resilience of this species in Indian conditions and the quality of its milk in terms of how it is processed into products, the planners should not become complacent with the increased share of milk production from crossbred cows and should instead place more emphasis on the production of milk and milk products from buffaloes. After crossbred, buffalo milk offers the second-highest returns on investment. Even in the final lactations, this profitability persists.

Indiscriminate breeding in field

India's history with animal breeding is a lengthy and complicated one. The livestock industry now has a fresh possibility as a promising food industry due to the high demand on agricultural land and growing human population. Low livestock productivity in India is a result of inadequate nutrition, a lack of planned breeding programmes, and the animals' interconnectedness with many socioeconomic problems. Breeding policy formulation must be done taking into account of regional demands. India forbids the slaughter of cattle, so a legal strategy is needed to prevent widespread indiscriminate mating, which may result in significant financial and infrastructure costs for castrating stray males and bolstering cow rehabilitation facilities. Without affecting the already-existing PT for exotic cattle breeds used in local cross-breeding programmes, and PT of new genotypes obtained from crosses of exotic and local breeds of cattle, as well as for purebred buffaloes, genetic evaluation of indigenous cattle with progeny testing (PT) requires significant financial support. Small ruminants require special consideration due to their socioeconomic significance in rural India. The attention should be paid to the diminishing grazing supplies should go hand in hand with genetic development of small ruminants, which would require both strong political will and financial backing. Since animals are rarely sold on a weight basis, farmers are discouraged from adopting explicit breeding objectives like improving growth rate, which would increase the reach of breeding programmes for small ruminants. There is also a lack of linkage between the market and producers. In addition to government institutions, participation from the commercial sector, non-governmental organizations, neighborhood cooperatives, self-help organizations, and self-sustaining community-based breeding programmes can increase market links. To meet the breeding goals, it is crucial to strengthen the current infrastructure and add technical resources and qualified labor.

Milk Processing and Value Addition of Milk

By adding or separating liquid whole milk, value-added goods are modified or improved in some way. Cheese, paneer, ghee, yoghurt, and probiotic drinks are examples of value-added products. These goods are created by modifying and enriching them through the addition of additives, microbial fermentation, or simple mixing with liquid milk. With the right approaches, value addition can be a terrific tool for doubling the farmer's revenue and increasing the profit on liquid milk. It can be used as a strategy to increase farmers' propensity towards the dairy sector. The market for value-added functional dairy foods has expanded as a result of people's greater awareness of proper nutrition and diets in recent years.

Reason for value addition of milk

There are many justifications for turning milk into useful dairy products, including the following:

- Unlike fresh milk, dairy products have a longer shelf life.

- Fresh milk storage is a problem, as it is bulkier and more perishable in nature.
- There is a meager market for raw milk.
- In comparison to fresh milk, preserved value-added functional food products can be carried over longer distances without spoiling, which could result in greater financial gains.

Challenges in value addition of milk

- Value-added functional dairy products can be a fantastic tool for ensuring that milk and milk products provide the greatest possible health benefits. In order to add value to fresh or raw milk, a medium- or large-scale dairy farm, or a collecting centre with a steady supply of raw milk from a group of farms is must.
- Lack of understanding among producers, processors, and consumers regarding the significance of value addition
- Irregular availability of clean, high-quality milk for processing; and insufficient financial incentives for dairy farmers and business owners to build infrastructure for processing at the farm, village, and regional levels
- Lack of infrastructure for effective marketing of the finished product.
- Lack of training and short courses offered to dairy farmers and business owners on a regular basis.

A range of nutritional components found in milk, many of which have great functional and nutritional attributes, can be exploited to create dairy products with added value. Milk is a nutrient-rich liquid. Adding value to milk will benefit society in many ways in terms of health while also boosting the economies of dairy producers and farmers. If value addition is designed, processed, and sold wisely, it can generate foreign exchange for the nation.

Challenges in the Indian Dairy Industry

- Despite the rise of the dairy sector, India still has problems with poor milk quality, low production, and a lack of adequate infrastructure. The following are a few difficulties:
- Large numbers of animals, not productivity, were typically responsible for Indian dairy's success. A strong need exists for superior animal genetics, breeding facilities, and cutting-edge breeding techniques including artificial insemination and embryo transfer, among others.

- India can concentrate on producing specialty goods made from buffalo milk, such as Mozzarella cheese that are suited to the demands of the intended customers.
- Green fodder and high-quality feed are in acute and escalating shortage. A substantial need for high-quality feed and fodder to meet the dietary needs of milking animals is being created by the growing trend of high-producing animals.
- Diagnostic tools for animal diseases are necessary, as well as quality healthcare. Animals with high yields require special care, and as time goes on, this tendency is propelling the animal health market.
- Insufficient chilling plants and bulk coolers are needed to prevent contamination and spoilage at the village level.
- At milk collection centres, adequate infrastructure for quality testing is also needed. The issue is made worse by a shortage of qualified personnel to conduct quality testing. Consumer demand for safe food is growing quickly, which is creating a lot of potential.

Strategies for future

Improvement projects for native breeds that have the potential to contribute to and be a part of the future production system should be on priority, started or strengthened in light of the scenario of a changing climate. Indigenous cattle breeds including the Gir, Red Sindhi, Sahiwal, Kankrej, and Rathi have potential. Mega semen stations should be built in order to meet the demands of semen production. Delivering AI services at doorstep is essential for obtaining a 40–50% conception rate. It is vital to create a workforce that is properly trained and to offer mobility and incentives. Large breeding businesses ought to be supported and given the task of providing AI services. The need for AI professionals should be determined, and those hired should be taught at reputable institutions. AnGR preservation ought to be carried out. Priority should be given to endangered breeds that possess distinctive features. Each breed must have at least one farm in the area where it was first domesticated. Immunization campaigns should be conducted for FMD, PPR, Brucellosis, and other serious diseases. There should be the establishment of an extensive national network of diagnostic laboratories. Breed development programmes ought to incorporate sexed semen, embryo transfer, and ovum pick-up technologies.

3

Dairy Business Challenges

All across the world, the dairy industry is facing numerous problems and changes. Dairy development has received a considerable boost from factors including access to technical advancements, improvements in infrastructural facilities, and the development of modern marketing insights. It is interesting to see how the dairy sector is changing globally. For rural India, this sector has a significant socioeconomic benefit. By taking advantage of the substantial unorganized market base in the Indian dairy industry, this benefit can be further increased. The exploration of sustainable dairy farming practices is also of importance, particularly in light of opportunities for market expansion for the dairy industry and small, marginal, and landless farmers. The chapter would cover sector-specific challenges, advantages, rising demand, and prospects for nation in the future.

Producers of milk, processors of milk, marketers, retailers, and consumers make up the majority of the dairy industry's stakeholders. In order for the dairy industry to grow, milk producers, the value chain, and consumers are all equally important. Major nations in this field are observed to pay attention in accordance with their priorities. It has been discovered that numerous nations, including Denmark, Netherlands, Australia, China, and many more, place a high importance on milk producers. Israel, India, Brazil, and many other nations have followed the example and are now developing dairy products in a "Value Chain based" manner. However, a number of these nations, including Sweden, Italy, and others, have embraced "customer based focus." However, Germany is perhaps the only nation that has taken into account all the aspects, including the value chain, customers, and milk producers, in their evaluation for development of dairy sector.

Milk producers must incorporate developing technologies and better farm management techniques in order to meet modern demands. The productivity of farms is increased by this technology and management techniques. However, producers receive subsidies in many industrialized nations, including the USA, Mexico, and others, which promotes bulk manufacturing. In these nations, export subsidies are provided in addition to tariff and non-tariff barriers. Because of this, it is difficult for producers in developing nations to compete with the low import prices from wealthy nations. Poor small-scale farmers in developing nations like Kenya, the Dominican Republic, Jamaica, and others are dealing with the reality on a daily basis. Exporters from the European Union were able to penetrate and capture markets in developing nations by offering lower prices. The decision by the World Trade Organization to stop export subsidies is good news for farmers and consumers in emerging and impoverished nations. Additionally, cost reduction, reduced capital expenditure, and increased milk production are some of the typical methods that aid small and medium herds in operating more profitably. Productivity and managerial competence work together to build a sustainable firm.

Consumer demand for protein-rich, healthful, and "clean label" milk products is rising. Consumers today are looking for "functional foods." Particularly among athletes, older citizens, infants, and toddlers, there is an increasing trend in the demand for milk protein through dairy-based products that have been created. To create more value for customers, local marketers must find and use consumer insights. Consumer trends are constantly evolving, and modern eating habits frequently include things like protein-rich foods, frequent eating out, and nutritious snacks. Online purchasing is also on the rise. Dairy products are in high demand because they are an affordable source of protein, have a pleasant taste, and are simple to obtain. In fact, businesses must use data from databanks to target customers with specific offers. Even in smaller establishments, retailers must manage their inventory just-in-time to balance supply and demand.

Farmers' Milk Selling Price

It is well known that the existence of vendors and middlemen in the supply chain prevents milk farmers from receiving the fair price for their product. Milk cooperatives must step forward in this area. This issue can be mitigated by reducing the unorganized dairy sector. Extension services and structured marketing facilities can help the situation even further. Cost reduction in milk production and processing is necessary for farmers to survive. Value-added product production will increase the profitability of dairy business owners.

Dairy producers can make more money by producing low- and high-fat milk for consumption by general consumers and by targeted consumers, respectively. When farmers sell their produce directly to consumers, they can benefit from higher profit margins. It suggests that when there is no cooperative infrastructure, the middleman's power becomes more apparent. Middleman takes advantage of milk farmers in need. The profitability and viability of the relevant businesses are significantly impacted by the cost structure of milk production, breed diversity, animal maintenance, and other factors. These characteristics require enough care and ongoing development.

The size of the farm, the processing capacity, the quality of the milk, and the cost of producing it are all important factors in the global market. The popularity of milk and milk products has increased as there is a greater desire for healthier products. Full-cream milk, standardized milk, toned milk, and skimmed milk have all been created using different ratios of fat and solids-non-fat to identify their uses and predicted market worth. Additionally, value-added goods like ice cream, ghee, etc. can provide higher profit margins. Population growth, rising incomes, shifting food consumption patterns, and better income elasticity for dairy products have all contributed to rising milk demand or opening up new prospects for the sector's expansion.

In India, milk output is predicted to increase at a rate of 4.4% during the next few years. However, it is predicted that milk demand will increase by 7% annually. Due to the increasing spending flexibility of low income rural and urban families, the demand for milk will continue to increase during the upcoming years. Despite being the world's greatest milk producer, India only accounts for a tiny portion of worldwide trade. Exporting ethnic sweets and ready-to-eat milk products is something that Indian businesses are interested in doing. Although it has created an export inspection system, sanitary and phytosanitary concerns prevent it from exporting.

Challenges Faced by Dairy Sector

Fragmented Supply Chain: Maintaining quality and quantity within a diverse supply base is dairy's primary problem. Dairy requires more intricate supply chain operations and logistics due to its perishable nature in order to maintain freshness and safety. The industry also sees tactics like adulteration and excessive use of antibiotics to increase production.

Price Sensitivity: Even little hike can have a significant impact on milk producers. For instance, even minor changes in consumer employment and income can have a big effect on milk demand.

Unorganized Nature: Contrary to farmers who grow sugarcane, wheat, and rice, the bulk of cattle farmers are not organized. This prevents the development of clout to promote their rights.

Data Deficiency: The cost of producing milk is not officially estimated or updated on a regular basis. Even still, the value of milk production in India exceeds the value of all the rice and wheat produced.

Poor returns: In contrast to important agricultural commodities in the nation, such as wheat and rice, milk does not have a minimum support price. Dairy cooperatives are also not the favoured option for small producers or those without access to land. The cooperatives use a fat-based pricing strategy that is 20–30% less expensive than the open market price. Dairy cooperatives also purchase over 75% of milk in its lower price range.

Competition from alternatives: Some people avoid traditional cow and buffalo milk in favour of healthier substitutes like "Soy Milk" or "Almond Milk." They contend that compared to conventional dairy products, plant-based milk products have a significantly lower carbon footprint.

Solution to the challenges

Start-ups: The government ought to assist new businesses that develop a methodical strategy. For instance, the dairy technology business Country Delight delivers high-quality milk to customers' doorsteps. The necessary assistance for the producers to reach value-added markets including ice cream, yoghurt, cheese, and whey should be provided. These divisions display 20% profit margins, which is significantly greater than the 3-5% margin for straightforward milk products. A reliable market and a profitable price for the milk must be provided to the dairy farmers. The quantity of milk should be given more weight for determining price than the milk's fat content. The hub-and-spoke design should be the government's main priority. Under this, all integrated milking, feed production, and milk processing facilities should be located on the main farm. A fundamental infrastructure for managing the livestock and milking should be present on the connected farms. For inclusive development, the hub should also offer their spokes technical, veterinary, and training support.

Increasing Productivity: The output of animals must be increased and breeding facilities, health care, and dairy animal management must all be improved. This may lower the price of making milk. Additionally, by assuring the accessibility of veterinary services, artificial insemination (AI), nutrition, and farmer education, milk production and productivity can be increased. Both the government and the dairy business can be quite helpful in this regard.

Augmenting Production, Processing and Marketing Infrastructure: It is essential to have a production, processing, and marketing infrastructure that can match international quality standards if India is to become a dairy exporting nation. Therefore, a comprehensive plan for producing high-quality and secure dairy products is required. This plan should be developed with the necessary required support. Investments in solar-powered dairy processing machines are also necessary to overcome the lack of infrastructure in rural areas and the power scarcity. Dairy cooperatives also need to be strengthened. The government ought to support farmer-producer organizations in this endeavour.

The dairy industry must grow sustainably given the existing circumstances. The environmental, dietary, and socioeconomic requirements of the nation should be met by this development. The government should take decisive action to increase the profitability of dairy production for small and marginalized farmers.

Main challenges in promoting exports of dairy products from India

Despite extraordinary expansion in milk production to become the biggest milk producing country in the world, dairy exports from India face a number of challenges that may be summarized as follows:

- Regardless of being the largest milk producer in terms of total quantity, India's average milk yield per cattle remains much inferior compared to developed and even many other developing countries.
- The little size of milch-animal holdings in India makes it hard to adopt mechanized system of milking, cooling and chilled storage which hampers the efforts to get better worth at the farm production stage.
- India being an enormous milk consumer owing not only to its large population size but also due to the major vegetarian population in the world whose only supply of animal based essential nutrient is milk, much low surplus is left for exports unlike other key dairy exporting countries.
- In several developed counties India faces a view of being a country with common incidence of foot and mouth disease (FMD) despite the random incidences of the disease in some part of the country. India needs to make determined efforts both to eliminate FMD and augment its view to be free of any the disease in milch-animals.
- Cow milk is the only trendy milk in the majority developed countries whereas India produces major quantity of buffalo milk. As overseas buyers are not always sure of appropriateness of buffalo milk for human utilization, they frequently be adamant upon dairy products manufactured from cow milk.

Approach to encourage Dairy Exports from India

As substantial transportation and logistics remains a vital worry for exports of dairy products from India, an incorporated strategy for overall development of export logistics in terms of creating cold chain facilities for shipping and storage needs to be adopted. In addition, India needs to concentrate upon exports of value added products with increased shelf-life and better packaging to compete in international markets. Rigorous efforts to market particularly in building global brands and establishing international marketing channels are also needed.

India needs to tackle successfully the emerging challenges under the new trade order affecting exports of dairy products. Furthermore, as import tariffs have significantly declined and quota restrictions fast vanishing in international markets, there is a strong alarm that high income countries are increasingly making use of quality standards as a frightening obstacle to dairy exports from India and other developing countries. The research institutions and scientists in India need to keep a close vigil on such compulsory standard specifications in international markets so as to defeat the newly emerging international trade barriers.

4

Standards for Milk and Milk Products

In India, the Ministry of Health and Family Welfare (MOH&FW), dealing with the subject of food quality and safety at the national level, has been designated the nodal ministry for maintaining food standards. The instrument for ensuring food quality and safety at the national level is through a legislation titled Prevention of Food Adulteration Act, 1954 (PFA) and Rules made there under as amended from time-to-time.

BUTTER

PFA Standards

- Milk Fat - not less than 80% by weight
- Common Salt - not more than 3% by weight
- Curd - not more than 1.5% by weight
- Diacetyl - 4 ppm
- Moisture - not more than 16%

Microbiological Standards (BIS)

Yeast & Mold count/ml	Quantity
< 20	Good
21-50	Fair
51-100	Poor
> 100	Very Poor

Coliform count should not be more than 10/ ml for satisfactory quality butter.

GHEE

AGMARK Standards

S.NO.	Tests	All-India	Winter	Summer
1.	Baudouin	Negative	Negative	
2	Phytosterol acetate	Negative	Negative	
3.	B.R. reading (40 ºC)	40.0-43.0	41.5-44.0	42.5 – 45.0
4.	R.M. Value	Not less than 28.0	Not less than 23.0	Not less than 21.0
5.	P Value	1.0-2.0	0.5-1.2	0.5-1.0
6.	Moisture (%)	Not more than 0.3		
7.	Free fatty acid (% oleic)			
a.	Special grade Agmark Red label	Not more than 1.4		
b.	General Grade Agmark Green label	Not more than 2.5		

R.M. Value in Tamil Nadu: 24

ICE CREAM

BIS Standards

Chemical

Characteristics	Requirements
Weight (g/Litre) (min)	525
Total solids (% wt) (min)	36.0
Milk fat (%wt) (min)	10.0
Acidity (% lactic acid) (max)	0.25
Stabiliser / emulsifier (%wt) (max)	0.5
Sucrose (%wt) (max)	15.0

Microbiological

Test	Limit
Standard plate count (per g)	not more than 2,50,000
Coliform count (per g)	not more than 90
Phosphatase	negative

MILK POWDER

BIS Standards

Chemical

Characteristics	Whole Milk Powder	Skim Milk Powder
Flavour and odour	Good	Good
Moisture (%wt.) (max)	4.0	5.0
Total milk solids (%wt.)	96.0	95.0
Solubility		
a. Solubility index (max) ml	15.0 (if roller dried)	15.0 (if roller dried)
2.0 (if spray dried)	2.0 (if spray dried)	
b. Solubility (%wt.) (min)	85.0 ml (if roller dried)	85.0 ml (if roller dried)
98.5 ml (if spray dried)	98.5 ml (if spray dried)	
Total ash (on dry basis) % wt (max)	7.3	9.3
Fat (% wt)	Not less than 26	Not more than 1.5
Titratable acidity (% lactic)	1.2	1.5

Microbiological standards

Types	WMP & Extra Grade SMP	Standard Grade SMP
Total bacterial count, max, cfu/g	40,000	50,000
Coliform count	Absent in 0.1 g	Absent in 0.1 g
Salmonella	Absent in 25 g	Not specified
Staph. aureus (coagulase positive)	Absent in 0.1 g	Not specified
Shigella	Absent in 25 g	Not specified

CHEESE

BIS Standards

Characteristics	Natural Cheese (hard variety)	Processed cheese	Processes cheese spread	Cheese Powder
1. Moisture % by mass. maximum	43	47	60	5
2. Milk fat (on dry basis) % by mass. minimum	42	40	40	40
3.Salt (added NaCl) % by mass maximum	3	3	3	5.5
4. Coliform count, per gram maximum	-	-	-	10
5. Stabilizer /	0.1%	4%		
6. Sorbic acid / Nisin	0.1%	0.1%		

Suggested bacteriological standards

S.No.	Product	Yeast and Mold Count	Coliform	Aerobic spore count	Anaerobic spore count
1.	Processed cheese/gm	60	Nil	200	100
2.	Processed cheese spread/gm	60	Nil	200	100

CONCENTRATED MILK

BIS Standards

Characteristics	Full Cream	Skim Milk
Total milk solids (% by wt., minimum)	31.0	26
Fat (% by wt.)	min 9.0	max 0.5
Sucrose (% by wt., minimum)	40	40
Acidity (% lactic acid, maximum)	0.35	0.35
Bacterial count (cfu/g, maximum)	500	500
Test for Coliforms	Negative	Negative
Yeast and Mold count (cfu/g, maximum)	10	10

(*Source*: http://ecoursesonline.iasri.res.in/mod/page/view.php?id=65131)

5

Dairy Animals in India

The world's greatest source of cow and buffalo diversity is India, and countries like Australia, South Africa, Latin America, and the USA employ Indian cattle breeds to grow their livestock industries. The number of cattle in the nation represents 17% of all cattle in the world. 22-25% of the Indian cow population is made up of the best native milk, draught, and dual-purpose animal genetics. The majority of local cattle breeds are excellent drafters. Native livestock breeds are clearly superior in their use of subpar feed, have better heat tolerance, and greater resistance to tropical diseases. 53 different cattle breeds and 20 recognized buffalo breeds are mentioned in Indian literature. The genetic diversity of cattle and buffalo is being increased around the world through the usage of these diverse and wealthy breeds. Since the beginning of time, livestock have played a crucial role in Indian agriculture and in the rural economy. Over one-fourth of the agricultural GNP is contributed by livestock, which also employs around 9% of the agricultural labour force and is increasingly regarded as a food source. Since 2000, the Government of India has been carrying out the "National Project for Cattle and Buffalo Breeding (NPCBB)" with an emphasis on genetic upgrading of cattle and streamlining AI services and support systems in an effort to boost milk production. Due to the limitations of a small herd size, and lack of initiative to support breed societies, progress in the area of bull production and assessment has been gradual. Additionally, focus on maintaining local milk breeds did not leave much of an impression.

The National Dairy Plan (NDP), a project funded by the World Bank that is scheduled to begin in early 2012, aims to increase the production of dairy animals by offering high-quality semen, on-demand AI services, and scientific feeding. With 1/5th of the global bovine population, India is ranked number 1 in this category. In India, nondescript animals make up about 70% of the total cattle population and 60% of the buffalo population. Additionally, the vast majority of descript cattle are from breeds used for both draught and dual purposes. In the discussion below, registered breeds of cattle and buffalo are described.

Cattle Breeds of India

Utility based classification of Indian cattle

Milch breeds

- Red Sindhi
- Sahiwal
- Gir

Dual purpose

- Tharparkar
- Kankrej,
- Dangi
- Rathi
- Deoni
- Hariana
- Ongole

Draught Breeds – Four Types

- Short horned : - Nagori, Bachur
- Lyre horned : - Malvi, Kherigarh
- Himalayan Region - Ponwar, Siri
- Mysore type : Prominent fore head with long and pointed horns which rise close together - Hallikar, Umbalachery, Alambadi, Pulikulam, Amritmahal, Burgur, Khillari, Kangayam

Followings are the important cattle breed of India:

1. Amritmahal

Breeding Tract: Karnataka

Use: Work - Transport and Draught

Well-known draught breed known for its power and endurance. Animals are very active. Bullocks are particularly suited for trotting and quick transportation.

Color: Grey, but varies from white to almost black.

Horns are long. Emerge from the top of the poll fairly close together in backward and upward direction

Characteristics: Long head tapering towards muzzle; long horns.

2. Bachaur

Breeding Tract: Bihar

Use: Work - Draught

Bullocks can work for lengthy periods without any break.

Color: White or Grey

Horns are small and stumpy, curving outward, upward and downward.

Characteristics: Medium sized compact animals with straight back. Forehead is flat or somewhat convex.

3. Bargur

Breeding Tract: Tamilnadu

Use: Work - Draught

Animals are very perceptive and fiery in disposition and are hard to train. They are light in built and are developed mainly for carrying out agricultural operations in the uneven and hilly terrain. Cattle of this breed are unmatched in speed and endurance in trotting.

Color: Brown with white markings.

Horns are closer at root and are inclined backward, outward and upward with a forward curve; sharp at the tip.

Characteristics: Brown color with white markings and light brown color of horns.

4. Binjharpuri

Breeding Tract: Odisha

Use: Food-Milk; Work- Draught; Manure

Dual purpose breed

Color: White. Some animals are Grey, Black or Brown in color

Horns are curved upward and inward.

Characteristics: Medium sized, strong dual type animal. Hump, neck, and some region of face and back are black in color irrespective of coat color in males

5. Dangi

Breeding Tract: Gujarat, Maharashtra

Use: Work - Draught; Food - milk

Dangi cattle are broadly used for ploughing, harrowing and other field operations, and also for carting timber from forest area. The breed is well known for its outstanding working qualities in heavy rainfall areas, rice fields and hilly tracts.

Color: Dangi cattle have distinct white coat color with red or black spots distributed unevenly over the body.

Horns are short and thick with lateral pointing tips

Characteristics: Uneven distribution of black or red spots over the body. Forehead is slightly protruding.

6. Deoni

Breeding Tract: Karnataka, Maharashtra

Use: Work - Draught, Transport

Deoni bullocks are favored for heavy work.

Color: Body color is usually spotted black and white.

Horn emerges from the side of the poll behind and above the eyes in out ward and upward direction.

Characteristics: Black and white spotted body color, drooping ears, prominent and slightly bulging forehead

7. Gaolao

Breeding Tract: Madhya Pradesh, Maharashtra

Use: Work - Transport; Food - Milk

This is a quick trotting type breed suitable for quick transportation in hilly areas.

Color: Blackish white in males and White in female.

Horns are short and stumpy and curve slightly backward.

Characteristics: Head noticeably long and narrow tapering towards muzzle. Eyes almond shaped and placed slightly at angles.

8. Ghumusari

Breeding Tract: Odisha

Main Use: Work-Draught; Food-Milk; Manure and fuel

Basically a draft breed

Color: Mainly White sometimes Grey

Horns are mainly White sometimes Grey

Visible Characteristics: Small sized, strong and draft type animal. Small head with forehead being flat, broad and depressed in between the eyes

9. Gir

Breeding Tract: Gujarat

Use: Food - Milk; Work - Draught

Gir cows are good milk producers. Bullocks can pull heavy loads.

Color: pure red though some are speckled red.

Horns are peculiarly curved. Starting at the base of the coronet they take a sideways descending and backward curve and again incline a small upward and forward taking a spiral inward sweep, ending in a fine taper- therefore giving a half moon appearance.

Characteristics: Long and pendulous ears folded like a leaf. Ears hang all the time and their inside face forward.

Age at first calving (Avg. Months): 46.08

Av. Milk yield per lactation (kg): 2110

Peculiarity: Gir is a world-renowned breed known for its tolerance to stress conditions. it has the capacity for yielding more milk with less feeding and is resistant to various tropical diseases.

10. Hallikar

Breeding Tract: Karnataka

Use: Work - Draught

Best draught breed of Southern India.

Color: White to light grey with darker shadings on shoulder and hind quarters.

Horns are emerging near to each other from top of poll and are carried upward and backward, ending up with pointed tips.

Characteristics: White markings or irregular patches round the eyes and cheeks, and neck or shoulder region are the distinctive features of this breed

11. Hariana

Breeding Tract: Haryana

Use: Food - Milk; Work - Draught and Transport

Hariana is a prominent dual purpose breed of north India - widely spread in Indo Gangetic plains. It is primarily reared for bullock production. Cows are also fairly good milkers.

Color: Animals are white or light grey in color. In bulls color in between fore and hind quarters is relatively dark or dark grey.

Horns are small in size.

Characteristics: White color, long and narrow face, well marked bony prominence at the centre of poll and small horns.

Age at first Parturition (Avg.Months): 51.3

Parturition Interval (Months): 15.88

Milk yield per lactation (kg): 997

12. Kangayam

Breeding Tract: Tamilnadu

Use: Work - Draught and Transport

Bullocks have a good ability for work and are used for all agricultural operations.

Color: Coat is red at birth, but changes to grey at about 6 month of age. Bulls are grey with dark color in hump, fore and hind quarters, face and legs.

Horn are long and strong, take backward, outward and upward sweep and then curving inward with tips tending to meet each other to form crescent shape.

Characteristics: Long horns nearly completing a circle, long hump and strong body.

13. Kankrej

Breeding Tract: Gujarat, Rajasthan

Use: Work - Draught and Transport, Milk

Agricultural operations and road transport in village area are mainly carried out by bullocks of this breed.

Color: Varies from silver-grey to iron grey or steel grey. In males fore & hind quarters and hump are somewhat darker than the rest of the body.

Horns are sturdy and curved outward and upward in a lyre shaped fashion. They are curved with skin to a longer distance as compared to other breeds.

Characteristics: Heaviest breed of cattle. large pendulous and open ears.

Milk yield per lactation (kg): 1738

Peculiarity: The gait of the Kankrej is unusual to the breed; the head is held noticeably high, the stride is long and even and the hind hoof is placed well ahead of the impression of fore hoof. This gait is called 1¼ paces (Swai chal, Proud gait, Majestic gait).

14. Kenkatha

Breeding Tract: Madhya Pradesh, Uttar Pradesh

Use: Work - Draught

Very popular for light draught on road and for cultivation.

Color: Color varies from grey on the barrel to dark grey on the rest of body.

Horns are emerging from the outer angles of the poll in markedly forward direction and terminate in sharp points.

Characteristics: Small, strong and fairly powerful animal. Horns directed forwards.

Peculiarity: They thrive on poor feed and fodder resources.

15. Khariar

Breeding Tract: Odisha

Use: Work-Draught; Food-Milk; Manure and fuel

Basically drafts breed. Used for draft power in its native tract.

Color: Mainly Brown sometimes Grey

Horns are straight, upward and inward.

Characteristics: Small sized, sturdy draft type animal. Hump, neck, and some region of face and back are dark in color

16. Kherigarh

Breeding Tract: Uttar Pradesh

Use: Work - Draught

Bullocks are very good for draught purposes.

Color: Animals have white coat color.

Horns are upstanding, curving outward and upward. These are thick at the base. Horn formation is typical of lyre-horned. Horns are medium in size Characteristics: Small but active animal.

Peculiarity: Animals are resistant to diseases and the spending on treatment is almost nil.

17. Khillar

Breeding Tract: Karnataka, Maharashtra

Use: Work - Draught

Bullocks are known for quick draught capability.

Color: Males are dark over the fore and hind quarters with unusual grey and white mottled marking on face.

Horns are long and pointed and follow the backward curve of the forehead ending in pointing tips.

Characteristics: Bow shape, long horns.

Peculiarity: Bullocks are extremely valued as fast powerful draught animals. They can travel miles without showing any signs of exhaustion.

18. Kosali

Breeding Tract: Chhattisgarh

Use: Manure, Draught and Milk

Color: Mainly light red

Horns are stumpy and straight

Characteristics: Muzzle, eyelids, tail switch and hooves are black. Head is broad, flat and straight. Udder small and bowl shaped.

19. Krishna Valley

Breeding Tract: Karnataka, Maharashtra

Use: Work- Draught

This is a heavy draught breed and used exclusively in black cotton soil of the watershed of river Krishna.

Color: The common color is grey-white with a darker shade on fore and hind quarters in males.

Horns are curved and usually emerge in an outward direction from the outer angles of the poll curving somewhat upward and inward.

Characteristics: Massive body, distinct bulge in the forehead.

20. Malnad Gidda

Breeding Tract: Karnataka

Use: Draught, Milk and Manure

They play a most important role in the rural economy of this region by providing milk, manure and draft power with negligible inputs.

Color: The main coat color is black with light shades of fawn on thigh and shoulder region.

Horns are generally small and straight.

Characteristics: They are small in size with compact body frame weighing around 80-120 kg. Tail switch-black, hump-small, udder small and bowl shaped.

21. Malvi

Breeding Tract: Madhya Pradesh

Use: Work- Draught and Transport

Well known for rapid transportation, stamina and ability to carry heavy load on rough roads.

Color: white or white grayish in males, with neck, shoulders, hump and quarters almost black.

Horns are curved, emerge from the outer angle of the poll in an outward and upward direction.

Characteristics: Strong well built whitish grey animal with lyre shaped horns.

22. Mewati

Breeding Tract: Haryana, Rajasthan, Uttar Pradesh

Use: Work- Draught

Mewati cattle are potent and docile, and are useful for profound ploughing, carting and drawing water from deep wells.

Color: usually white with neck, shoulders and quarters of a darker shade.

Horns: Outwards, upwards, inwards in majority of animals; Tips pointed.

Characteristics: Face is long and narrow with straight, sometimes somewhat bulging forehead.

23. Motu

Breeding Tract: Odisha

Use: Work- Draught; Manure: Food-Milk

Color: Mainly Brown (Reddish) sometime Grey.

Horns Shape and Size: Straight, upward with rounded tip.

Characteristics: Small sized, strong and draft type cattle. Animals are typically polled and brown in color.

Peculiarity: Cow dung is considered as significant product from rearing of these cattle

24. Nagori

Breeding Tract: Rajasthan

Use: Work-Draught and Transportation

Color: usually white or light grey

Horns emerge from the outer angles of poll.

Characteristics: white, very alert and lively animal with long and narrow face.

Peculiarity: Bullocks are large and powerful. They are capable of heavy draught work. There is a tendency to legginess and lightness of bone, though feet are strong. This characteristic has given the breed its agility.

25. Nimari

Breeding Tract: Madhya Pradesh

Use: Work- Draught

Used for all agricultural operations as well as for transportation.

Color: Brownish red (copper color) and white giving spotted appearance

Horns usually emerge in a backward direction from the outer angles of the poll

26. Ongole

Breeding Tract: Andhra Pradesh

Use: Work- Draught; Food- Milk

Color: Ongole have a glossy white coat. Males have dark grey marking on head, neck and hump, black points on knees and pasterns, black muzzle and black eye-lashes.

Horns are short and stumpy, growing outward and backward from the outer angles of the poll, thick at the base and firm without cracks.

Characteristics: stumpy horns, large fan shaped and fleshy dewlap serrated with smooth flowing folds instead of narrow constrictions.

Peculiarity: The Nellore or Ongole breed hardiness, disease resistance and its capacity to thrive on scanty and dry fodder have been quite successfully exploited for development and upgrading their local stock by many countries. Ongole breed known for heat tolerance and for disease resistance.

27. Ponwar

Breeding Tract: Uttar Pradesh

Use: Work- Draught and Transport

Bullocks are quick mover, good for agricultural operations.

Color: usually brown or black with white patches.

Horns emerge outward and then curve inward with pointed tips. Horns are average in size.

Peculiarity: Animal behaves semi-wild type and is very difficult to handle.

28. Pulikulam

Breeding Tract: Tamilnadu

Use: Draught, Game and Manure

Bulls are used in Bull riding (Jallikattu).

Color: Dark grey in males and white or grey in females

Horns Curved-outwards, upwards, backwards and inwards ending with pointed tips, tips are wide apart.

Characteristics: Small size, compact body and short legs. Muzzle, eyelids, tail switch and hooves are black. Fore Head is wide and level with groove at centre. Hump large in males and small in females.

29. Punganur

Breeding Tract: Andhra Pradesh

Use: Work- Draught and Transportation; Food- Milk

These animals are used for agricultural operations. Bullocks are used for carts for races.

Color: Punganur cattle are white, grey or light brown to dark brown or red.

Horns are crescent shaped and often loose curving backward and forward in males and lateral and forward in females.

Characteristics: Short stature.

Peculiarity: Dwarf cattle

30. Rathi

Breeding Tract: Rajasthan

Use: Food- Milk

Color: The animals are usually brown with white patches all over the body. The lower body parts are generally lighter in color as compared to rest of body.

Horns are curving outward, upward and inward.

Characteristics: Face is broad between eyes and slightly dished.

Dewlap is fine and loose.

Tail – long, fine tapering to a good black or white switch well below the hock.

Age at first calving (Av. Months): 46.4

Calving Interval (Months) 17.07

Milk yield per lactation (Kg): 1560

31. Red Kandhari

Breeding Tract: Maharashtra

Use: Work – Draught and transport

Bullocks are used for heavy work.

Color: The color is uniform deep dark red, but variations from a dull red to almost brown are also found.

Horns are evenly curved and medium sized.

32. Red Sindhi

Breeding Tract: The original breeding tract is in Pakistan but some herds are present in Orissa, Tamil Nadu, Bihar, Kerala and Assam states of India.

Use: Food- Milk

Color: This breed has distinctly red color. Red shades vary from dark red to dim yellow.

Horns are thick at the base and emerge laterally and curve upward.

Age at First Calving (Av. Months): 43.54

Calving Interval (Months): 14.57

Milk yield per lactation (kg): 1840

33. Sahiwal

Breeding Tract: Punjab, Rajasthan

Use: Food- Milk

One of the best dairy breed of zebu cattle. It is being utilized for improvement of local stock.

Color: Brownish red color, shades vary from mahogany red brown to more grayish red. Extremities in bulls are darker than rest of the body color.

Horns are stumpy and short to medium.

Characteristics: Pale red color, short horns and loose skin.

Age at first Calving (av. Months): 41.7

Calving Interval (Months): 15.6

Milk yield per lactation (kg): 2325

34. Siri

Breeding Tract: West Bengal, Sikkim

Use: Work- Draught

Bullocks are the only source of draught power.

Color: Animals are either black with white patches or brown with white patches.

Horns are curved outward, forward and slightly upward with sharp and pointed tips, and are medium in size.

Characteristics: Color pattern alike to that of Holstein Friesian.

35. Tharparkar

Breeding Tract: Gujarat, Rajasthan

Use: Food- Milk; work- Draught

Color: Animals are white or light grey. Face and extremities are of a darker shade than the body. In bulls neck, hump, and fore and hind quarters are also dark.

Horns are set well apart gradually upward in the same line as that of the poll, with blunt points inclined inward. In males, horns are thicker and short.

Characteristics: Convex forehead.

Age at first calving (Av Month): 41.03

Calving Interval (Month): 14.18

Milk yield per lactation (kg): 1749

36. Umblichery

Breeding Tract: Tamil Nadu

Use: Work- Draught

These are light built draught animals developed for work in marshy paddy fields.

Color: Red at birth which changes to grey at three to four months of age. Bulls are deep grey with black extremities.

Horns are curving outward and inward and occasionally spreading laterally. These are wide in bulls and thin in cows. Horns are very small in size.

Characteristics: White marking on face, limbs and tail. All the legs below hocks have white marks either as Socks or as Stocking. Even a portion of hooves is white.

37. Vechur

Breeding Tract: Kerala

Use: Food- Milk; Dung- Manure

Color: Animals are light red, black or fawn and white.

Horns are small, thin curving forward and downward. In few cases they are very small and are hardly visible.

Characteristics: Extremely small sized /miniature cattle.

Recently registered cattle breed

38. Belahi

Cattle of the Belahi breed are also referred to as Morni/Desi. The breeding area for this breed is located in the Shivalik foothills of Haryana State and includes Chandigarh, Ambala, Panchkula, and Yamunanagar. The term "Belaha" is used to describe color combinations. This breed of animals migrates and requires fewer resources to sustain.

The most prevalent animal colors are reddish brown, grey, or white. Horns have an upward and inwardly curling sickle form. The body color pattern of animals is uniform but unique, and they are medium in size. Whiteness can be noticed in varying degrees on the ventral region of the body, as well as the face and extremities. Head is straight, broad with a prominent poll.

39. Gangatiri

The breeding tract of Gangatiri encompasses the Uttar Pradesh districts of Varanasi, Mirzapur, Ghazipur, and Ballia as well as the Bihar district of Bhojpur. This breed's animals produce medium amounts of milk and have good draftability. Completely white or grey is the color. The medium-sized horns start outward and curling upward and inward behind and above the eyes before emerging from the side of the poll and terminating with pointy points. The forehead has a deep groove in the centre and is prominent, straight, and broad. Hooves, the tail switch, and the eyelids are all typically black in color.

Lactation produces approximately 1050 kg of milk on average, with variations between 900 and 1200 kg and an average fat content of 4.9%.

40. Badri

Due to their prevalence in Uttarakhand's steep terrain, the Badri breed—also known as "Pahadi"—is a dual-purpose animal. Animals have a balanced gait when walking and are well adapted to hilly terrain and climatic circumstances. Black, brown, or grey are the most common colors, with white being rare in females. Horns have an upward and inward curve. Badri is a little, vivacious, and agile. The forehead has a medium to large hump and a noticeable poll. Badri cows produce 632 kg of milk on average lactation, with an average milk fat content of 4%.

41. Lakhimi

The entire state of Assam is home to the Lakhimi breed of cattle, which serves two purposes. Animals can survive in climates that are subtropical. Bullocks are excellent carrying and ploughing animals, especially in the muddy fields used for growing paddy. Animals are small, horned, and have relatively short legs. They are also brown or grey in color. The backline is somewhat curved, and the hump is medium in size. Udder is formed like a little bowel. It has straight horns. The average milk supply each lactation is only 359 kg, and the average milk fat is 5.3%.

42. Ladakhi

Ladakhi is a breed of cattle native to Jammu and Kashmir that originated in the Ladakh region. This breed can survive at great altitudes, in extremely cold climates, and in hypoxic environments. Compact body and short legs make it better suited to mountainous terrains. The animal's primary colors are black and brown. The horns' pointy tips are located over the forehead and are typically curled, pointing slightly forward and upward. Animals are little in stature and have short legs. They also have a small hump. Every day, a cow produces between 2 and 5 kg of milk. Manure, milk, and draught are the main uses for animals.

43. Konkan Kapila

The Konkan Kapila, sometimes referred to as "Konkan gidda" and "Konkan," is a Maharashtrian draught breed. The farmers can employ the animals for a variety of agricultural tasks. This breed can subsist on grazing and is well adapted to low input production systems. Animals come in a range of colors, with reddish brown and black being the two most common. Horns usually have a straight shape; emerge from the poll's side behind and above the eyes, moving upward and backward until ending with sharp tips. Animals have compact bodies that range in size from tiny to medium, horizontal ears, and a straight forehead. The Konkan Kapila cow's low milk production, good draught ability, suitability for its native region's mountainous terrain and hot, muggy climate are the major contribution to the livelihood of small and marginal farmers.

44. PodaThurpu Cattle

These compact, medium-sized cattle are found in Telangana's Nagarkurnool area. Animals either have a reddish-brown coat with white spots or a white coat with brown patches. The base of the horns is large, and they are oriented straight up or forward. Most often, the forehead is convex with a deep groove in the middle. Cattle from the PodaThurpu region are raised mostly for draught. Bullocks are popular in both dryland and wetland agriculture because they have great draught capability in terms of endurance, speed, and stamina. The daily milk yield is between 2 and 3 kg, while the lactation milk yield is between 494 and 646 kg.

45. Nari Cattle

Gujarat and Rajasthan are the native home of nari cattle. These have good migratory abilities and are medium in size. The majority of cows have white or greyish white coats, whereas bulls typically have white, greyish white, or black coats. Horns have a spirally curved shape and point forward or outward. Most often, the forehead is broad and slightly concave. These cattle serve two purposes: producing milk and pulling draught. Milk production varies from 5 to 9 kg per day and 1119 to 2223 kg per lactation. Excellent in draft ability.

46. Dagri Cattle

Gujarat's Dahod, Chhotaudepur, and portions of Mahisagar, Panchmahals, and Narmada districts all have Dagri cattle populations. The color of the coat is primarily white, occasionally with a touch of grey. Animal is small, compact, and with a straight forehead. Horns can be straight with sharp points or upwards curved in the shape of a lyre. Breed is widely used as a draught animal for farming operations in difficult terrain. Cow produces 1.5–3 kg milk per day and average 650 kg per lactation and is mostly used for family use.

47. Thutho

Cattle of the Thutho breed are also referred to as "Ameshi," "Sheapi," "Chokru," and "Tseso." There are thutho cattle accessible in all of Nagaland's districts. This breed is mostly utilized for draught and manure. Animals can feed on hill slopes even during rain since they are well adapted to mountainous areas. The animals are typically black or brown, with occasional white patches on their faces and bodies. Animals are amiable, hardy, medium in size, and well constructed. Straight and tiny is the forehead. Horns have an upward and outward curve, stumpy and short. Typically, cows are not milked.

48. Shweta Kapila Cattle

Goa State's North Goa and South Goa districts are home to Shweta Kapila cattle. From the muzzle to the tail switch, everything is white, including the eyelashes and the muzzle (whitish brown). Animal of short to medium height, cylindrical bodies with a straight face. Horns are small, straight that point upward and outward, and small to medium hump. Small to medium sized and bowl-shaped udders have spherical teats. The average daily milk yield is 2.8 kg, while lactation milk yields range from 250 to 650 kg.

49. Himachali Pahari Cattle

The Himachali Pahari cattle population inhabits the mid- to high-hills of Himachal Pradesh and has adapted to the mountainous terrain, bitterly cold climate, and lack of forage. Small to medium-sized animal with a predominantly black and blackish brown coloration, a compact cylindrical body, short legs, a medium hump, ears that are positioned horizontally, and a somewhat long tail. Horns are typically modest in size and curled mostly laterally and upward. Well suited as a source of milk, draught, and manure for mountainous livestock production systems. Male adults can weigh 200 to 280 kg, while female adults can weigh 140 to 230 kg. 1 to 3 kg of milk can be produced daily, and 350 to 650 kg of milk can be produced during lactation.

50. Purnea Cattle

This cattle breed is found in the Bihar districts of Araria, Purnea, and Katihar, as well as the neighboring regions of Kishanganj, Supaul, and Madhepura, Purnea. These creatures are small and have three distinct coat colors: grey, red, and black. These are raised primarily for their milk, dung, and to a lesser measure, draught power. These cattle have a medium hump, a small to medium naval flap, a small to medium sized udder, and a small to medium sized dewlap. Head is medium in size. While some Purnea cattle have relatively prominent polls, the majority of them have flat polls. Horns are often carried upwards, straight, and occasionally laterally. The range of milk production each day is 1 to 5 kg, and lactation milk production is 360 to 785 kg.

51. Kathani Cattle

This breed is from Maharashtra. The most common body colors observed are white, red and blackish. The color of horn was generally grey and blackish. Hump small to moderate in cows and fairly developed in bulls. Dewlap was small to medium in size. Body was compact, cylindrical in shape with short legs and animals were noticed to be strong and active. Ears were short, erect and horizontal in orientation.

52. Sanchori Cattle

This breed is from Rajasthan. Body color is Silver grey in the female and iron grey in the male. The predominant coat colors of bull are grey but white, black and red may be there. Head is dished in centre. Face is Narrow. Eyes are bright showing great intelligence. Ears are Very long pendulous, drooping in majority otherwise horizontal orientation can also be observed.

53. Masilum Cattle

This breed is from Meghalaya.

Important crossbred cattle

Karan Swiss

Karan Swiss is a crossbred of cattle developed at the National Dairy Research Institute of India (NDRI) Karnal by crossing of Sahiwal and Brown Swiss breed cows

The breed is typically light gray to dark brown in color.

The share of Brown Swiss within the breed ranges from 1/2 to 3/4.

The Karan Swiss cows are giving good quantities of milk. High producing females are producing 5000 to 6000 kg with a 4.78% butterfat during lactation.

Karan Fries

The development of Karan Friesian (KF) has been done by cross-breeding with the Holstein Friesian breed of the Tharparkar breed found in Rajasthan.

This breed has been developed by the National Dairy Research Institute (NDRI) located in the state of Haryana (Karnal). The breed has black and white spots on the body, forehead and tail of the cow.

Karan Fries cows are considered as very good milk producer breeds. These cows give milk up to 10 litters to 15 liters per day, and they have the capacity to provide milk up to 4000 litters a year.

Sunandini

Sunandini is a composite breed of cattle developed in India by crossing nondescript cattle with Brown Swiss, Jersey cattle and Holstein Friesian cattle. Jersey and HF bulls from many parts of the world are extensively utilized to generate this breed.

Frieswal

The Frieswal cattle is developed by crossbreeding Holstein and Sahiwal in different Military Farms to cover all the agro-climatic zones of the country. The average milk production potential of Frieswal cows based on 300 day milk yield is more than 3300 kg,

Phule Triveni

Phule Triveni breed of cow is developed by Mahatma Phule Krishi vidyapeeth, Rahuri, Maharashtra. This breed consists of Holestian Friesian, Jersey and Gir combinations. It is cross bred and has average milk production of 3000 to 3500 litre per lactation with 3.8 to 4.2 % fat in milk.

Buffalo Breeds of India

Classification of Buffalo

The domestic water buffalo Bubalus bubalis, belongs to the family Bovidae, sub-family Bovinae, genus bubalis and species arni or wild Indian buffalo. Buffalo are classified into two distinct classes: swamp buffalo and river buffalo.

Swamp Buffaloes

The swamp species is most prevalent in Southeast Asia, where it is mostly used as a draught animal, notably in rice farming. Its name is derived from the swampy or marshy environment in which it naturally occurs. Swamp buffaloes share morphological traits with wild arni. Its potential as a meat producer has recently come to light. There are no separate breeds of the swamp variety, although some strains are larger than others. Due to their extremely poor milk production, swamp buffaloes are not employed for milk production. On a big scale, river and swamp buffalo crossings have been attempted in Thailand, the Philippines, Vietnam, and China. These crossbred buffaloes are strong working animal and yield more milk and high-quality meat than their native brethren.

River Buffaloes

The river buffalo is the most common type in India, Pakistan, Bulgaria, Hungary, Turkey, Italy and Egypt. The river buffaloes are massive in size mostly with curled horns, prefers to wallow in clean water and rivers, thereby the name. River buffaloes have been selected for milk production to a larger extent and some river breeds.

Followings are the important buffalo breed of India:

1. Banni

Breeding Tract: Gujarat

Use: Food- Milk

Color: Mainly Black, sometimes Copper

Horns are Medium to large, horns are vertical and upward in direction with inverted double/single coiling.

Age at first calving (Av. Months): 40.3

Calving interval (Months): 12.24

Milk yield per lactation (kg): 2857.2

2. Bhadawari

Synonym: Suargori

Breeding Tract: Madhya Pradesh, Uttar Pradesh

Use: Food-Milk; work- Draught

Color: Blackish copper to light copper. Color of legs is usually like wheat straw.

Horns are black curling slightly outward, downward before running backward parallel and close to neck and finally turning upward.

Characteristics: Two white lines "Chevron" or Kanthy are present on the lower side of the neck.

Age at first calving (Av. Months): 44.6

Calving interval (Months): 16

Milk yield per lactation (kg): 1294

Milk Fat (%): 7.88

Peculiarity: This breed is an proficient converter of coarse feed into butterfat and is known for its high butter fat content in the milk.

3. Chilika

Breeding Tract: Odisha

Use: Food- Milk; work- Draught; Manure

Color: Brownish black or black

Horns are curved upward, inward.

Characteristics: Medium sized with compact body, strong legs and small udder

4. Jaffarabadi

Breeding Tract: Gujarat

Main Use: Food- Milk; Work- Draught and Transport

Color: Black

Horns exhibit wide variation, but frequently emerge out by compressing the head, go downward sideways, then upward and inward finally forming a ring like structure.

Characteristics: Horn shape is a peculiar of this breed.

Age at first calving (Av. Months): 45

Calving Interval (Months): 14.6

Milk yield per lactation (kg): 2239

Peculiarity: Jaffarabadi is the heaviest Indian breeds of Buffaloes

5. Kalahandi

Breeding Tract: Odisha

Use: Milk, draught, manure

Color: blackish grey

Horns are horizontal going backwards, upward and inward, curved to make half circle appearance

Characteristics: Head convex, hump small, udder round and medium in size. Tail extends below hock with coarse hairs on switch.

6. Marathwadi

Breeding Tract: Maharashtra

Use: Food- Milk; Work- Draught and Transport

Color: Grayish black to jet black.

Horns are parallel to the neck, reaching upto shoulder but never beyond shoulder blade.

Characteristics: Horns reach up to shoulder not like in Pandharpuri buffaloes where these may reach up to pin bones.

7. Mehsana

Breeding Tract: Gujarat

Use: Food- Milk

These breed is considered to be a cross between Surti and Murrah

Color: Mostly black

Horns are generally sickle shaped. They are generally bent downward and then take a curve.

Characteristics: Eyes are very prominent, black and bright.

Peculiarity: The breed is known as constant milker and regular breeder.

8. Murrah

Breeding Tract: Haryana, Delhi

Use: Food- Milk and Meat; work- Draught

Males are used both for breeding as well as draught purpose.

Color: Jet Black

Horns are tightly curved in a spiral form and short in size.

Characteristics: Jet black color, tightly curled horns.

Age at first calving (Av. Months): 43.4

Calving interval (Months): 14.9

Peculiarity: This breed is found in almost all parts of the world and is being bred either in pure form or is being used for grading up local buffaloes.

9. Nagpuri

Breeding Tract: Maharashtra

Use: Food- Milk; Work- Draught

These buffaloes are used for heavy draught purpose.

Color: Black colored animals with white patches on face, legs and tail tips.

Horns are flat, curved and carried back on side of the neck nearly to shoulders like a pair of swords, horn tips pointed, upward in orientation mostly.

Peculiarity: Farmers maintain this breed mostly due to its very low maintenance cost, high fat in milk and regularity in breeding.

10. Nili Ravi (Panch Kalyani)

Breeding Tract: Punjab

Use: Food- Milk

Color: Normally black

Horns are tightly curved and circular in cross section, small in size.

Characteristics: Usually walled eyes (large bulging eye/light-colored iris) and white marking on forehead, face, muzzle, legs and tail. The most desired character of female is the possession of these white marking known as "Panch Kalyani".

Age at first calving (Av. Months): 45

Calving interval (Months): 16

Milk yield per lactation: 1850

Milk Fat (%): 6.8

Peculiarity: Similar to Murrah in almost all respect for some white markings on extremities and walled eyes (large bulging eye/light-colored iris).

11. Pandharpuri

Breeding Tract: Maharashtra

Use: Food- Milk; Work- Draught

Color: Usually black but varies from light to deep black.

Horns are sword shaped; Horns are very long and extend beyond shoulder blade, sometimes up to pin bones.

Peculiarity: It is resilient and well suited to dry conditions.

12. Surti

Breeding Tract: Gujarat

Main Use: Food- Milk

Color: Coat color varies from rusty brown through silver-grey to black.

Horns are flat and sickle shaped

Characteristics: Medium sized animal, two white bands below the neck and sickle shaped horns.

Peculiarity: The surti buffalo is lighter in body weight, as compared to heavy breeds, consume less feed, and thrives on limited or no green fodder and produce milk. It is popular with landless, small and marginal farmers.

13. Toda

Breeding Tract: Tamil Nadu

Use: Work- Draught; Socio-cultural-Religious ceremonies

Color: calf is usually fawn at birth. In growing calves, at about 2 months, the fawn color changes to ash grey. In adult buffaloes the predominant coat colors are fawn and ash-grey.

Horns long and are variable in shape. Horns are thick at the base.

Characteristics: A narrow band of thick hair covering the top line from the crest of neck to the point of origin of tail, two chevron marking- one just approximately the jowl and other anterior to the brisket.

Recently registered buffalo breeds

14. Luit (Swamp)

This breed is also known as "Assamese Swamp". Breeding tract includes Jorhat, Sibsagar, Lakimpur, Dibrugarh, Tinsukia, Dhemaji, Golaghat, Majuli and Biswanath districts of Assam.

Mainly used for milk and draught. Color of buffalo is mainly black.

These swamp buffaloes are medium sized strongly built animals with prominent wither and short tail.

The animals are characterized by broad and concave fore head with prominent eyes and wide muzzle and distinct semi-circular horns in both sexes.

Horns are curved laterally backward and then upward forming a semi-circle.

The average lactation yield of Luit buffalo is 449 kg with an average milk fat of 8.68 %.

15. Bargur

These buffaloes are found in the Bargur hills in Tamil Nadu. Coat colors vary from black to light brown or brownish-black.

Greyish white stockings from carpal/tarsal joint to fetlock are present predominantly in females.

The milk yield of the animals ranges from 1.5 to 2.0 liters per day and mainly used for household consumption.

16. Chhattisgarhi

These buffaloes are distributed throughout the Chhattisgarh state.

Coat color is black.

Animals are medium built with proportionate body.

Horns are medium to large in size and directed laterally backwards and then upwards with pointing tips.

Milk yield ranges from 3 to 6 kg/day.

17. Gojri Buffalo

In Punjab as well as the Kangra and Chamba districts of Himachal Pradesh, the Gujjar population raises Gojri buffaloes using a semi-migratory management style. These buffaloes are typically brown or black in color and have proportionate, medium-sized bodies. The majority of the medium-sized horns are bent to form a large loop. They are used to generate draught power and milk. Milk production varies from 3 to 8 kg per day and 800 to 1200 kg during lactation.

18. Dharwadi buffalo

Karnataka receives Dharwadi buffalo distribution. Dharwadi is a medium-sized buffalo. A black coat is found. The ears are up. Teats are cylindrical and udder is average in size. It is raised primarily for milk production. The typical lactation production is 972 litres of milk. A daily milk production of 1.5 to 8.7 litres is possible. The milk is used to make the well-known Dharwad Peda. The animals are suitable for regions with little rainfall.

19. Manda buffalo

The Manda buffalo is found throughout Odisha. It is a strong buffalo that is ideally suited to the plateau of the Koraput region of Odisha and the high ranges of the Eastern Ghats. Ash and grey are the predominant body colors, while the hair is copper in color. Leg's lower portion is lighter. Broad, slightly lateral, spreading backward and inward, and forming a half circle are the characteristics of the horns. It is raised for milk, manure, and draught. The amount of milk produced each day varies from 1.2 to 3.7 kg. Most Manda buffaloes are raised using a low input system.

20. Purnathadi buffalo

The Purnathadi buffalo is found in Maharashtra. Purnathadi is a medium sized buffalo, docile in nature with a typical whitish to brown coat color.

Population Statistics of dairy animals

Cattle breeds

One of the main contributors to the livestock population in the nation is cattle. Cattle make up 36.04 percent of the total animal population, according to the 20th Livestock Census. The two breed categories Exotic/Crossbred and Indigenous / Nondescript can be used to categorize the species. Animals that fall under these broad categories are further categorized in the 20th Livestock Census into the corresponding breeds recognized by the National Bureau of Animal Genetic Resources (NBAGR).

Exotic and Crossbred Cattle

One of the high-producing groups of cattle is the exotic and crossbreed category. Exotic animals are defined as those with origins outside of one's own country. The Jersey and Holstein Friesian breeds of cattle are the most major foreign cattle in India (HF). Crossbred animals are those that are created by mating native species with exotic breeds. Other exotic crossbred animals are those created by mating native species with foreign breeds other than Jersey and Holstein Friesian (HF). Exotic and crossbred animals make up about 26.5% of the entire cattle herd, according to the 20th Livestock Census. Breeds such as the Jersey, Crossbred Jersey, Holstein Friesian (HF), Crossbred HF, and other Exotic crossbred are categorized under exotic and crossbred.

Indigenous Cattle

Indigenous animals are those that come from Descript/Non-descript breeds of indigenous heritage. According to the 20th Livestock Census 73.5% of cattle are indigenous. Breeds that fall within the category of "graded breed" of that particular indigenous breed have more than 50% of the phenotypic traits of that breed. In addition to the Non-Descript animals, ICAR-NBAGR, Karnal has recognized 53 indigenous breeds of cattle in the nation.

Breed of Buffalo

In this nation, buffaloes are the second-largest producer of milk and the third-largest contributor to the population of livestock. According to the 20th Cattle Census, buffaloes make up 20.47% of all livestock. In addition to the Non-Descript animals, ICAR-NBAGR, Karnal has recognized indigenous 20 breeds of buffalo in the nation. 20.3% of the total number of buffaloes is pure breeds, followed by 34.3% of graded buffaloes and 45.4% of nondescript buffaloes.

Goat breeds

The second-largest species of livestock and contribute in the production of milk after cattle and buffaloes, goats play a vital role in the small ruminant group. Goats make up 27.74% of the animal population, according to the 20th Livestock Census. Indigenous goat populations total 148.88 million, of which 27.4% are pure breeds, 9.1% are graded breeds, and the remaining 63.5% are nondescript breeds. In addition to the Non-Descript animals, ICAR-NBAGR, Karnal has recognized 34 indigenous breeds of goats in the nation.

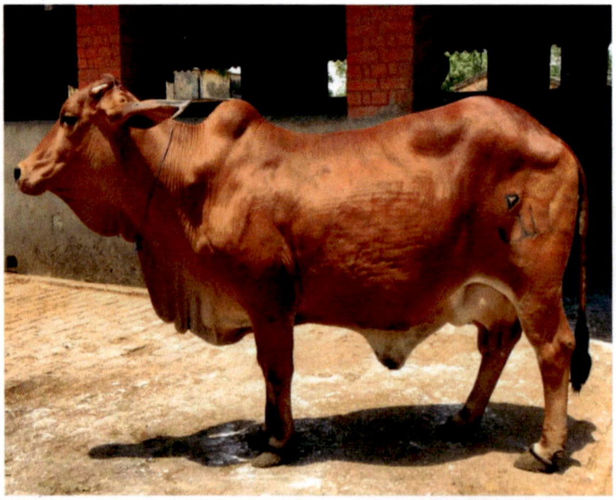

Fig. 1: Sahiwal cow (Source: DUVASU, Mathura website)

Fig. 2: Sahiwal cow herd (Source: DUVASU, Mathura website)

Fig. 3: Hariana cattle cow (Source: DUVASU, Mathura website)

Fig. 4: Murrah buffalo herd (Source: DUVASU, Mathura website)

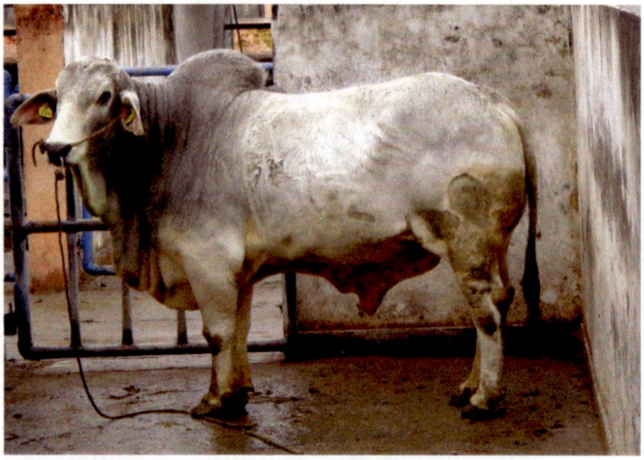

Fig. 5: Tharparkar cattle Bull

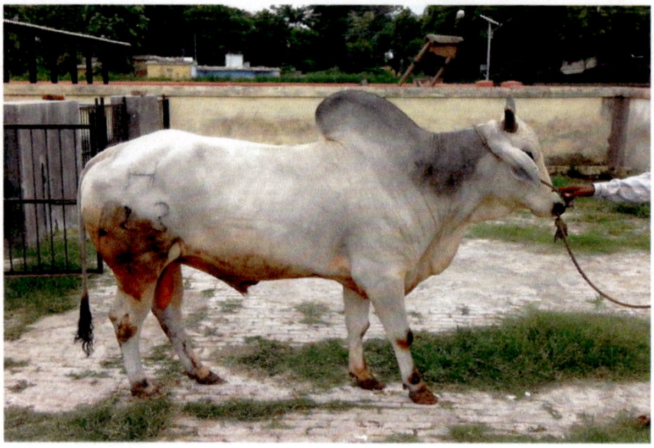

Fig. 6: Hariana cattle Bull

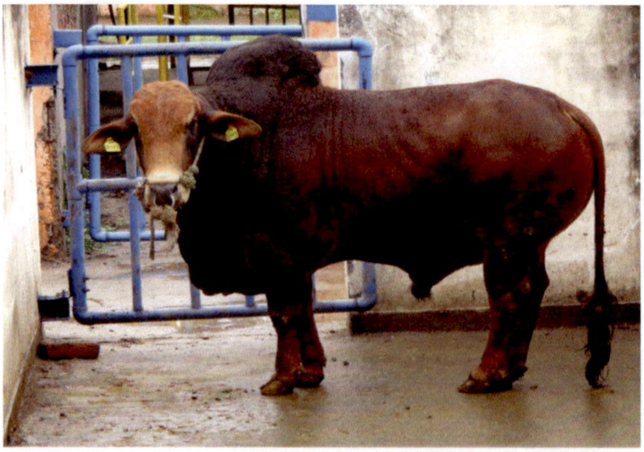

Fig. 7: Sahiwal cattle Bull

Fig. 8: Karan Fries cattle Bull

Fig. 9: Murrah buffalo Bull

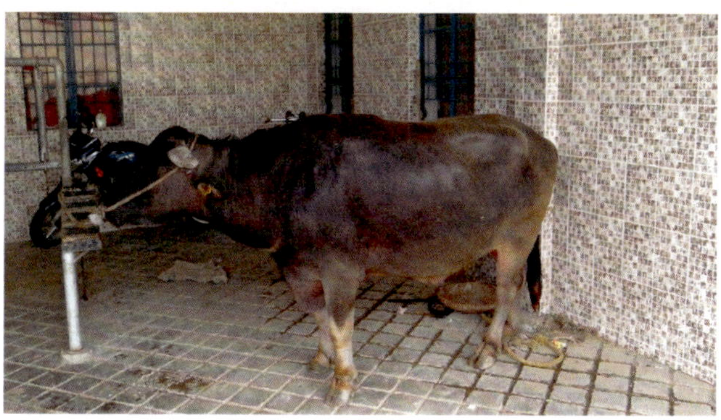

Fig. 10: Bhadawari buffalo Bull

Figure 11: Black Bengal Goat

Fig. 12: Black Bengal buck (Source: DUVASU, Mathura website)

Fig. 13: Barbari Goat

Fig. 14: Barbari buck

Fig. 15: Sirohi buck (Source: DUVASU, Mathura website)

Fig. 16: Jamunapari goat (Source: DUVASU, Mathura website)

Fig. 17: Jamunapari buck (Source: DUVASU, Mathura website)

Fig. 18: Beetal goats (Source: DUVASU, Mathura website)

6

Buffalo is the Future

India is home to an enormous number of buffalo. According to the 20th Livestock Census, more than 109 million buffaloes, or more than 56% of the world's population, are living in India. With 20 recognized breeds, it also has a very diverse collection of buffalo genetics. One of the most well-known buffalo breeds in the world, the Murrah, is also found in India. The buffalo is a versatile animal that can be used for milk, meat, and draught. India is the world leader in this industry, producing 76% of the world's buffalo milk. Over 187 million tones of milk were produced in India overall in 2018–19. According to official statistics, 74 million cows generated close to 90 million tones, while more than 51 million buffalo produced over 92 million tones. The average milk output of buffaloes is higher than that of cows. Additionally, milk is of higher quality. Traditionally, the amount of fat in anything is what determines its quality. Cow milk has an average fat content of 4.4%, compared to 6.7% for buffalo milk. The buffalo is not a common source of milk in the majority of the world. The majority of the packaged milk that is sold in the nation is a blend of cow and buffalo milk. All of the conventional sweets are also. But the buffalo is not a topic of much public discussion. Contrarily, the cow is the centre of attention.

The cow has no advantage over the buffalo in terms of basic dairy economics. A buffalo is a low-cost and low-maintenance animal for a dairy farmer. It is not picky or particular about the flavour or quality of its food. On coarse feed and crop leftovers, it flourishes and even survives. The buffalo is the best animal for small farms. They find it difficult to care for and manage pedigreed cows that require expensive feed. The buffalo continues to be a valuable asset even after it has finished producing milk since it can be sold. It is not restricted by any religious, cultural, or societal taboos, unlike the cow. It's common practice to sell off aged buffaloes and bull calves for meat. About 30% of the nation's total meat production comes from buffalo. The majority of it is exported. More than 1.2 million tones of products made from buffalo meat worth 251.7 billion rupees were shipped in 2018–19. While there are more than 5 million stray cattle in the country, which is an issue, there aren't any known stray buffaloes. With the advent of farm machinery, oxen are no longer needed to plough fields. Therefore, the issue of runaway cattle is expected to worsen. In order to control this issue, the country would need to spend a significant amount of money on sustaining unproductive animals. Given the stringent prohibition on killing cows, the buffalo is a superior choice for both farmers and consumers. The nutritional value of a certain type of milk frequently has little scientific support. Because cows and buffaloes were domesticated for dairying purposes many millennia ago, their milk is nearly identical and ideally suited for human use.

They have been developed to produce more milk in greater quantities over time, and today's consumers find their milk to be quite palatable. Milk contains between 83% and 91% water. Milk from buffalo and cows differs somewhat yet significantly. In comparison to cow milk, buffalo milk contains a lot more solids. As a result, more dairy products can be produced and processed from it. However, nothing created from buffalo milk can be made from anything made from cow milk. The yield of products made from buffalo milk, such as cream, butter, paneer, or condensed milk, is also higher. Buffalo milk cheese offers superior chewing and stringing qualities. Yogurt from Buffalo is thicker. Butter and buffalo ghee offer great textures. Cow milk also has advantages. Traditional Indian soft sweets like chhena, sandesh, rosgulla, and chum chum are excellently appropriate for it. In general, buffalo milk is more adaptable. Athletes prefer it in particular for its higher calorific content. It is a healthier option for kids because of the higher level of calcium and other minerals. The buffalo needs more respect and focus due to its many benefits. This indigenous species is a national treasure.

7

Genetic Improvement of Animals

Dairy farmers are faced with a number of fundamental inquiries. The first inquires as to what dairy animal he needs to keep. Is the breed with the best production performance the one that has adapted to the local environment the most? What animal is the best, asks the second? Is it the one with a show-winning conformation or the one with a strong ancestry? Is the ideal dairy cow one that produces the most milk, has the best feet, legs, and udder support, or combines the performance of these characteristics in the best possible way? How can you breed animals so that their offspring would, if not be the best, at least be superior to those of today? Or to put it another way, how are animal populations genetically improved? Breeders have a lot to say about these, and nobody really has all the answers. The response that breeders choose will define the course of genetic change for breeding operations, breeds, and even species, hence the question is crucial.

Best is a vague concept. There is no perfect animal for every circumstance. An animal that thrives in one area might not be the greatest choice in another environment or under different conditions. It necessitates a thorough understanding of critical attributes and the relationships between performance in these traits and variables including the physical environment, management practices, costs, and prices. These factors differ depending on the species and breed, as well as the organization of the breeding industry and the position of the breeder within it. In animal breeding, our main focus is on genetically altering animal populations. Therefore, from a genetic perspective, we want to know both the most desired genotypes and the most desirable phenotypes. That is because an animal's genotype provides the genetic background for its phenotypes and it is the genetic material that is passed on from parents to its offspring. Summarized in an equation: $P = G + E$, where P represents an individual's phenotype, G represents its genotype, and E represents the environmental effects- the effects that non-genetic factors have on an animal's performance. In other words, an animal's phenotype is determined by its genotype and the environment it experiences. There are various ways to use the word genotype. The entire set of genes and gene combinations that influence the variety of qualities that we are interested in can collectively be referred to as an animal's genotype. This section will utilize the genotype known as "tropically adapted" as an example. The genotype in this instance consists of all the genes and gene combinations influencing features like heat tolerance, parasite resistance, and other characteristics of tropical adaption. This chapter normally assumes that the word genotype is used in this manner. Another term for the set of genes and gene combinations that affect a certain trait in an animal is its genotype. Or, as we will learn later in this course, we can further narrow the definition of genotype such that it only relates to a certain gene. We can alter the genotypes of our animals' progeny through breeding techniques. Positive genotypic changes lead to enhanced phenotypes.

We must identify the features that are most crucial and the genotypes that are ideal for those attributes in order to answer the question, "What is the best animal?" If they have some expertise, most breeders have an opinion on the most important qualities and the best genotypes. The ideal animal, in the opinion of a breeder of Thoroughbreds, might be "... quick, yet with enough endurance and heart for the longer distances, and readily rated." "...a healthy pig with a strong growth and good carcass quality," might be the pig breeder version. There are almost as many of these opinions as there are breeders, and most of them are highly personal. Understanding the function of the genotype in the agricultural system is crucial for developing a more objective understanding of the critical features and optimum genotypes. This means that the physical environment in which animals are maintained, the management system, as well as economic circumstances, will all affect how important certain features are. If you give it some thought, it will become obvious that the system's many parts will interact with one another. The optimum preventive health programme, for instance, is determined by the types of diseases present in the region (the physical environment) and the price of vaccines, dewormers, etc. You need to be aware of alternative programmes, local pathogens, treatment costs, and how treatment programmes interact with these other elements to affect profitability in order to decide which health programme is the most cost-effective. Similar to this, the best genotype is influenced by the regional environment, the management techniques being used, and the input costs and animal product pricing. Understanding how environmental, managerial, and economic factors interact with the genotype to affect profitability will help you choose the ideal genotype. To put it another way, if we want to design rational goals for breeding programmes, or suitable breeding objectives, we need to be aware of the function of the animal and the connections between the genotype and other system components. Animal populations and subsequent generations of animals are improved by breeding, not the individual animals themselves, in order to improve their genetic makeup. Breeders employ selection and mating as their two primary instruments for this endeavour. We choose which individuals become parents, how many progenies they can have, and how long they will stay in the breeding population through selection. We choose which of the bulls we have chosen will mate with which of the cows we have chosen before mating. Animals' long-term genetic changes are induced through selection. Which animals become parents, how many progenies they can have, and how long they stay in the breeding population is all determined by the process. We're all familiar with the concept of natural selection. The main evolutionary factor behind genetic change in all living things is natural selection. Natural selection affects both wild and domestic species, contrary to what we typically imagine. Natural selection influences both wild animals and plants. For instance, all animals with fatal genetic flaws are naturally selected against; they never survive to have children. Although it cannot be denied, artificial selection is the type of selection that is most important in animal breeding. Simply said, selection works by allowing individuals with the finest gene combinations to procreate so that the offspring would, on average, have more desirable genes than the current generation of animals. The best breeding values are thought to be found in the animals with the best gene sets. From a genetic perspective, they are the animals who would make the best parents. When making a selection, we want to pick the animals with the best breeding potential—those that will pass on the best genes to the following generation. A population's future generations will be genetically improved as a result of successful selection as the proportion of desirable genes will gradually increase. Consider phenotypic selection, often known as mass selection, to understand how selection functions. The only data needed to make selection judgments in this type of selection is the candidate's performance. The animal's pedigree, the performance of its siblings, or any offspring it may have generated, are not taken into consideration. For instance, you would solely rely on the ewe lamb's own weaning weight when employing phenotypic selection for weaning weight to decide whether to keep the animal for breeding. After all, what are passed on from parent to child are the genes. In other words, it is presumable that the body size trait in mice is connected in some way to the breeding value of body size. Phenotypic selection for this attribute would be pointless if it weren't the case. Therefore, it is crucial to understand the relationship between phenotype and breeding value, which is demonstrated by heritability.

Phenotypic selection will work well to change the amount of a trait when heritability is high since phenotypes are typically strong indicators of underlying breeding values. Phenotypic selection is ineffective when heritability is low because phenotypes don't offer much information about breeding values. The mice's rapid growth in size suggests that body size is highly heritable. Not every attribute is as inherited. In animals, for instance, the heritability of fertility is typically fairly low. Statistical methods are used to calculate the degree to which relatives differ from unrelated animals in terms of the trait of interest when estimating the heritability of a trait. Most breeders of animals are unlikely to make selections based solely on data about each animal's performance. They will also make use of family members' information. For instance, a dog breeder who buys an eight-week-old puppy from another breeder usually won't make that decision only based on the conformation and personality traits that are readily apparent in such a young puppy. He wants to assess these qualities in the dam, the sire, and the other littermates. To understand more about the puppy's ancestors, he can request to obtain a copy of its extensive pedigree. Similar to this, when selecting a sire for artificial insemination (A.I.), cattle breeders consider more factors than only the sire's own growth rate. They are interested in learning more about his offspring's capacity for growth. Accuracy serves as a gauge for the level of the correlation between genuine breeding value and its prediction. Predictions of breeding values are typically accurate and accurate in that they closely reflect the variations in genuine breeding values of the animals being evaluated. We can perform well in selection because breeding value estimates are reliable. All of the traits addressed up to this point in this chapter, including milk output in dairy cattle, body size in mice, fertility, conformation, and personality in dogs, were polygenic features. Many genes affect polygenic features, and no single gene is thought to have an overriding influence. Segregation at numerous loci is the cause of the genetic diversity in these features. We simply knew there were many of them; up until recently, we understood little about the precise genes affecting these qualities. In order to describe the genotypes of animals while still being unable to identify specific genes, we must rely on phenotypic performances and breeding value projections. There are solid reasons to think that the size of the impacts of genes on any trait vary, with some having big effects and others having very modest ones. Advances in molecular biology now make it possible to pinpoint specific genes that influence quantitative traits. Programs for selection can make use of data on genetic markers connected to specific genes to increase the efficacy of selection. Following the identification of a specific gene, its biochemical and physiological functions can be investigated. Our knowledge of the nature of genetic variation in traits will significantly improve as a result of the findings of these investigations. Animal features are typically polygenic in nature. However, certain qualities are merely inherited, meaning that only one or a small number of genes are responsible for them. An excellent illustration is the horned/polled appearance of cattle with European ancestry. The presence or absence of horns on a cow is determined by a single gene. A significant percentage of single-gene illnesses are likewise regarded as serious issues even if they do not prevent affected animals from getting pregnant. The malignant hyperthermia syndrome in pigs and the hereditary eye condition in dogs are two well-known examples. Selection for simply inherited features differs from selection for typical polygenic traits since only a small number of genes influence these traits. We do not deal with breeding values, their predictions, or even the idea of heritability when dealing with merely inherited qualities. Instead, we just care to know whether an animal carries the precise allele or alleles of interest, and we base our choice of animals on that information. Effective selection against the condition is conceivable if it can be discovered before reproductive age, either through clinical evaluation or DNA testing. Pig breeders now have a far better chance of eradicating malignant hyperthermia syndrome from the population thanks to the discovery of the disease's gene in pigs and the subsequent creation of a DNA test. Due to the autosomal recessive nature of malignant hyperthermia in pigs, it is impossible to distinguish between the phenotypes of animals carrying two normal alleles and those carrying one defective allele. The DNA test's effectiveness hinges in its ability to identify carriers—animals heterozygous for the genetic disorder—before they reach reproductive maturity. The first of the two fundamental methods that animal breeders utilize to alter a species' genetic makeup is selection.

Mating is the second instrument. Which males are bred to which females is determined via mating system. It differs significantly from selection. When mating, you pair males and females from the specified group while selecting the group of animals you wish to be parents. Animals can reproduce in a variety of ways, and each way can be categorized by a system of mating laws. There are three reasons for using mating systems: (1) to produce offspring with extreme breeding value, (2) to make use of complementarity, and (3) to obtain hybrid vigour. By mating parents with exceptional breeding values, extreme phenotypes can be produced. One technique to create an animal of intermediate size is to breed huge animals with little ones. The offspring is optimal despite the fact that the parental genotypes are significantly different and neither is optimal. Crossbreeding, or the union of sires from one breed and dams from another, is demonstrated by the mating of a Charolais and an Angus. The Charolais and Angus breeding is, in fact, a complimentary one in crossbreeding, which is frequently employed to achieve breed complementarity. The progeny of crossbred cattle benefit from having both types of parents because Charolais are enormous French cattle noted for their quick growth and substantial muscling, while Angus are smaller British cattle known for their maternal capacity. For the purpose of creating hybrid vigour, or heterosis, these two breeds are also crossed. The performance of crossbred or hybrid animals is improved over that of pure-bred animals. To a greater or lesser extent, several traits exhibit hybrid vigour.

8

Health Management

Herd management practices in the handling, nutrition, milking techniques, sanitation, and housing of dairy animals play key roles in predisposing the individual animal as well as herds to diseases. The key factor influencing the reduction of health issues are the decisions made by herd managers or animal owners. In order to maximize output and profitability through disease control, herd management approaches in conjunction with a veterinarian programme can be most beneficial. Governments and international organizations are typically in charge of controlling diseases that are carried through vectors or wild animals, are transmissible from animals to humans, or are epizootics. National regulations typically mandate vaccination programmes against endemic illnesses, which dairy farmers, cooperatives, and dairy organizations must closely abide with.

Since there are more environmental pressures in the tropics, dairy farm management needs to be sophisticated. Small dairy cooperatives and smallholder farms, however, frequently lack the requisite managerial abilities. Basic education and skills are required for dairy farming itself. Dairy producers should also aware of basics of animal health just like veterinary assistants and doctors do. Public funds should be used to support vaccination programmes, production, and health services in order to promote the development of cattle and buffalo. Animal productivity may be impacted by climatic and environmental stress, particularly heat stress. Loss of hunger, a decrease in daily milk production, a rise in temperature, a rapid breathing rate, tongue protrusion, inability to lie down, and a propensity for remaining in areas with cool water, such are all indications of stress. The animal's owner should be able to see the symptoms and make any necessary adjustments to the surroundings and housing to ease the animal's stress and maximise comfort. Animals in good health are alert, engaged, have bright eyes without discharge, smooth, shiny skin, regular breathing and urination, and a moving tail that chases flies away. The maintaining of records of all animals and events pertaining to animals throughout their lives is a crucial tool for farm management. The government or dairy cooperatives provide, assist, and design record keeping in several nations. Individual cow cards should be used to keep records on farms, at the dairy health service, or at the artificial insemination service centre. The management, breeding, and progeny testing of dairy herds are the primary purposes of the records. Farmers can forecast future or preventative healthcare needs by using records of insemination, birth date, sire, dam, calving date, immunization date, health issues, treatment, milk supply, and nutrition. Additionally, they offer pertinent and useful data that helps veterinarians make accurate diagnosis. As a result, it is best to keep meticulous records for each animal, with forms that are simple to understand. Smallholders don't appear to place enough importance on maintaining accurate records. The dairy herd improvement programme is a useful tool for farm management, but the methods are frequently expensive and require numerous professionals to reach smallholders in the tropics. Teaching farmers to utilize and continue using data requires a lot of work. Tropical environments offer additional pressures that exacerbate health issues and raise the prevalence of disease. Good housing and care can help animals feel less stressed, which lowers their chance of being sick. Similarly, reducing the proportion of Bos taurus blood in smallholder crossbreeds lowers the demand for complex management and facilities.

Practical health care

Health care of dairy animals can be best described by age and state of the animal as set out in the following sections.

Newborn calf: First three days

- Use a dry towel or clean, dry hay to clean the newborn calf. This will increase blood flow and breathing.
- Clear the mouth and nose of slime to help with breathing, and while holding up the calf's back legs, let the head dangle down to let any water in the lungs, mouth, or nose out.
- To prevent local infection if the navel is too long, trim it, leaving two to three inches of the stalk attached, and then soak it in tincture of iodine. This surgery aids in the fast healing of the umbilicus and is crucial for preventing navel disease (omphalitis).
- One to two hours after delivery, feed the calf colostrum. The first six to eight hours are ideal for antibody absorption through the small intestine of calves. The calf should receive 10% to 15% of its body weight from the colostrum. To avoid an early infection, the calf must consume enough colostrum within the first 12 to 24 hours. The immune system of the dam produces antibodies IgG and IgM in the colostrum, which have

a high nutritional value and act as passive defenses against numerous infections.

- In general, following calving, the calf should be removed from the dam and brought to an isolated pen that is dry and clean. The straw used as bedding needs to be fresh, dry, and replaced frequently. When raising crossbred dairy cattle in some areas, the owner lets the calf nurse from the dam for the first three to four days before separating from it.

Calves from three days to one month

- Since milk replacer is less expensive than whole milk, it is used for routine feeding, which occurs twice daily. Commonly used bucket feeding should be thoroughly cleaned after each usage to prevent digestive issues brought on by unsanitary conditions.
- At around one week of age, teach the calf to consume concentrate and roughage. Food that is solid encourages rumen growth. Clean water must always be accessible in the pen.
- Calves are assigned numbers via tattoos or ear tags. In the initial weeks, extra-teats could be removed.
- Omphalitis (navel illness), scours (diarrhoea), pneumonia, and arthritis are common medical conditions during this time.

One month of age to weaning (3–4 months)

- When calves are one to two months old, they should be dehorned.
- Between three and eight months of age, all female calves should get the brucellosis (S19) vaccine.
- Weaning should occur when the calf is three to four months old, can consume more than one kilogramme of roughage and concentrate per day, or weighs between 80 and 90 kg (depending on the breed).
- Deworm the calf to protect it from flukes, tapeworms, and other internal parasites. Additionally, spray to get rid of external parasites like ticks.
- This time of year, parasites, bloating, and arthritis are issues to watch out for.

Calves 4–12 months of age

- Get vaccinated every six months for anthrax, hemorrhagic septicemia, and/or foot-and-mouth disease (FMD).

- By spraying, de-worm against internal parasites like roundworm, tapeworm, and flukes as well as get rid of external parasites like ticks.
- During this time, parasites, tick fever, pneumonia, diarrhoea, bloat, and arthritis are concerns to watch out for.

Heifers 12–18 months of age

- Keep track of the growth rate, which at first service should not be less than 300 kg for purebred cattle or 270 kg for crossbred calves.
- Perform a tuberculosis test and draw blood for brucellosis.
- Vaccinate every six months for hemorrhagic septicemia and FMD (foot-and-mouth disease).
- Deworming needs to be done every six months.
- Heat detection should be performed to establish the ideal timing for artificial insemination and the usage of chosen sperm in accordance with the region's or farm's breeding strategy.
- A veterinarian should examine heifers that need repeated insemination (more than three times).
- Heifers over 18 months old, over 270 kg, or both who have not displayed oestrus symptoms should be examined by a veterinarian.
- Every animal should have a pregnancy diagnosis performed 45 to 60 days following the previous insemination.
- Three-day sickness (also known as ephemeral fever), tick fever, and other infectious diseases and parasites are common health issues at this age.

Pregnant heifers – prepartum

- Feed pregnant heifers in poor condition with high-quality roughage and supplement with concentrate.
- For pregnant heifers, a mineral supplement can be used to avoid metabolic illnesses such milk fever.
- Vaccinate against diseases such as hemorrhagic septicemia, FMD, and others.
- Routine de-worming for both internal and external parasites is recommended. The cow may start acting restless and strays from the herd as part of the parturition process. Enlargement of the udder and

belly as well as discharge from the vulva are indicators that a calf has been born.

- During this time, it's important to watch out for abortions and heifers mastitis (mastitis prior to calving).

At parturition

- The owner must set up the calving area, which should be tidy, dry, peaceful, and remote in order to keep the pregnant animal close so that the owner can see and assist if it exhibits signs of trouble giving birth.
- The vulva will grow, the teats and udder will distend, the ligaments at the side of the tail-head will lose their elasticity, and the animal will become restless. Other signs include an obvious rise in mucus production, a liquefaction of the cervical seal, and an increase in the frequency of uterine and abdominal contractions.
- The cow will need assistance from a veterinarian if delivery is found to take longer than 24 hours and the allantoic sac has not protruded.
- Veterinary intervention will be required if the foetus has not expelled itself or if there haven't been any contractions for more than two hours following the rupture of the allantoic sac.
- There's a chance that milk fever, uterine prolapse, or a downer cow will manifest during this time.

After calving: seven days postpartum

- The foetal membrane should naturally expel three to eight hours after calving, or within twelve hours. The cow will need assistance from a veterinarian if the foetal membrane is retained for more than 12 hours.
- As soon as you can, milk the colostrum and give it to the calf.
- To lower the danger of infection by flies, remove the foetal membrane from the calving area or pen floor and clean the pen and the dam's rearing area.
- Feed the dam appetising food in good quality and quantity during this time because the dam may still be under the effects of labour and delivery stress.
- Mastitis, metritis, milk fever, uterine prolapse, and retained placenta are all possible during this time.

Milking

- Keep in mind that weight loss following delivery is a sign that your diet lacks appropriate energy.
- Attempt to manage and feed cows in groups according on how much milk they produce.
- Examine the reproductive system for metritis, ovarian function, and uterine involution one month after delivery.
- If the dam needs more than three inseminations, a veterinarian should examine her. The dam should exhibit oestrus symptoms no later than 60 days after giving birth.
- Mastitis, metritis, abomasal displacement, acidosis, and ketosis are hazards during this time.

Dry period

- Before allowing the animal to dry off, confirm the pregnancy once more.
- Drying-off should let the animal at least two months to recuperate and be ready to calve before the upcoming parturition. Mastitis prevention is crucial both before and after calving.
- Upholding the regular immunization programme.
- Treat parasites both internally and externally.
- Hoof trimming is possible at this time.
- Keeping up a good feed regimen during the dry season.

Replacement heifers

The animals that were purchased must have an accurate health history. Breeding certificates, individual cards from previous owners, and information on vaccines and health issues are all significant pieces of information. Before fresh animals to enter a farm, quarantine may be essential. Cattle should be treated as gently as possible when loading or unloading because shipping livestock causes stress. The quickest amount of time should be allotted for shipping, together with the coolest time of day. Before loading, the cattle's temperature should be taken if there is any uncertainty about their health. Instead of running the danger of stress-related disease or death, it is more cost-effective to treat feverish animals and postpone transportation.

Culling

- Small-holders prefer to keep every animal, even those that produce little milk or never get pregnant. It's essential to remove underperforming animals from the herd and replace them with better stock for effective herd management. Some smallholders' unwillingness to cull in accordance with best practices may be attributed to sentiment and diversion from other agricultural endeavours, which prevents them from paying attention to the productivity levels of individual cows. Serious dairy farmers will adopt the culling procedures.
- Farmers will either sell male calves for a low price or remove them from the herd as soon as possible because it is not economical to keep them. For future replacement animals, farmers prefer to maintain exclusively female calves. Compared to producing local multipurpose cattle.

Vaccination in Cattle and Buffaloes

The animal's health should be satisfactory at the time of vaccination. Avoid giving vaccinations to stressed animals (Like bad weather, Scarcity of fodder & water, Disease outbreaks, after transportation etc.) One to two weeks before to immunization, try to deworm animals and remove ectoparasites. After consulting with veterinary professionals, strictly adhere to the immunization regimen. Maintain vaccination records for the name of the vaccine manufacturer, the batch number, the expiration date, the dose, and the route of administration. Maintain the cold chain for vaccinations that may need to be stored at 2 to 8 degrees Celsius. After immunization, create a stress-free environment for the animals.

Table 8: Vaccination Schedules for Cattle and Buffaloes

Sr.No	Name of Disease	Age at first dose	Booster dose	Subsequent dose
1	Foot and Mouth Disease (FMD)	4 months and above	1 month after first dose	Six monthly
2	Haemorrhagic Septicaemia (HS)	6 months and above	-	Annually in endemic areas.
3	Black Quarter (BQ)	6 months and above	-	Annually in endemic areas.
4	Brucellosis	4-8 months of age (Only female calves)	-	Once in a lifetime

Sr.No	Name of Disease	Age at first dose	Booster dose	Subsequent dose
5	Theileriosis	3 months of age and above	-	Once in a lifetime. Only required for crossbred and exotic cattle.
6	Anthrax	4 months and above	-	Annually in endemic areas.
7	IBR	3 months and above	1 month after first dose	Six monthly (vaccine presently not produced in India)
8	Rabies (Post bite therapy only)	Immediately after suspected bite.	4th day	7, 14, 28 and 90 (optional) days after first dose.

Bovine diseases

The prevalence of animal diseases is a significant barrier to the expansion of the livestock industry. Losses from some of these illnesses, such as Foot & Mouth Disease (FMD), Brucellosis, etc., are enormous and frequently impossible to predict. FMD causes not only a decline in milk production and the sale of livestock products, but also infertility and a decline in the quality of the skins of the animals, including their ability to draw water. Thus, the commerce of milk and other cattle products is directly harmed by FMD. The most prevalent and significant illnesses in India include mastitis, foot-and-mouth disease, and brucellosis.

Mastitis

Mastitis is well known as a major issue because it results in serious wastage and undesirable milk quality. It is among the challenges of dairy development in the tropics, along with breeding improvement, nutrition management, and control of infectious, tick-borne, blood, and internal parasitic diseases. Mastitis was anticipated to be a significant illness in dairy cattle at the time modern dairy production was implemented in the tropics. The disease has brought disaster for the majority of tropical dairy producers today. Understanding the causes of mastitis and employing management strategies that prevent the infection from spreading are necessary for its control. Since the knowledge of dairy farmers is frequently limited, assistance from dairy scientists, extension agents, educators, and veterinarians is required. It is crucial that these assistants have sufficient understanding of mastitis management. There is a dearth of research on mastitis and how to treat it in tropical climates. The vast majority of the knowledge that currently supports the tropical dairy businesses comes from in-depth study in industrialized nations. The causal agents and infection pathways are lucky to be similar despite the environmental and management variances amongst different systems. Thus, there are several similarities between mastitis control strategies. The same programme that works in huge herds housed in pastured herds can also work in small herds in tropical developing nations and herds located in cold regions where cows are kept in barns for the majority of the year. When additional locally obtained specialized knowledge has been gathered through research, programmes may be adjusted to suit local needs. This section offers dairy farmers in the tropics useful advice on how to treat mastitis. A number of microorganisms, primarily bacteria, enter the mammary glands through the teat canal to induce mastitis, an infection of the udder. The cow's udder, as well as the floor, faeces, dirt, feeds, water, plant matter, and milking tools and equipment, are home to the microorganisms. Leukocytes travel from the bloodstream into milk in response to bacterial invasion and proliferation in order to combat the infection. This is the inflammatory response, which may go unnoticed in the form of subclinical mastitis or it may be severe enough to be categorized as clinical mastitis, which is characterized by pathological changes in the mammary tissue as well as physical, chemical, and typically bacterial changes in the milk. Chronic mastitis may develop if leukocytes are unable to keep the infection under control or if it is not treated to eliminate it. As long as the infection persists, such an infected quarter may lose up to 25% of its milk output and produce just low-quality milk. Mastitis can be brought on by more than one hundred different kinds of microorganisms. The majority of dairy producers are aware of clinical mastitis in their herds, but they are less aware of subclinical mastitis' prevalence and economic significance. Clinical mastitis is distinguishable by obvious abnormalities in the udder and/or milk and is simple to identify.

The subclinical type cannot be seen by looking at the cow or milk, but it can be found by doing tests to look for somatic cells in milk or for infecting microorganisms or other signs of inflammation. In a herd population, there will typically be 15 to 40 subclinical cases for every occurrence of clinical mastitis, and the majority of clinical cases are preceded by infection at the subclinical level. Long-lasting subclinical mastitis diminishes milk output, degrades milk quality, and frequently results in herd issues. Additionally, it is significant because it serves as a reservoir for mastitis bacteria that could potentially spread to other cows in the herd.

Significance of mastitis management

Mastitis has a negative impact on public health because it reduces milk production, alters the composition and quality of the milk, and causes financial loss. Mastitis can be effectively managed to help resolve these issues. The public health threat will be lessened, producers' incomes will increase, and milk production and quality will both improve.

- Production of high-quality milk is one of the primary goals of dairy farming, whether on large- or small-scale farms. This is required so that processors may sell milk of superior quality, which is desirable, and consumers can accept it. Consumers define good quality milk and milk products as being wholesome, attractive, predictable in flavour and taste, preserving their original nutritional properties, free of hazardous microorganisms and chemicals, and having a long shelf life. Using properly sanitized equipment and keeping milk at its best for appearance, flavour, nutritional value, and free from drugs and chemical residues, with the least amount of microbial contamination, producers must ensure that milk comes out of not only disease-free animals but also from healthy udders. Regardless of the causes, unhealthy udders that are primarily linked to mastitis generate milk that is undoubtedly of poor quality, either in terms of milk composition or bacterial contamination. Only effective mastitis management can enable farmers to control it, if not entirely mastitis-free, at least almost mastitis-free farms which will increase their profits.
- The evident sign of clinical mastitis is a decrease in milk production. The amount of inflammation, which may be determined by the somatic cell count in milk, determines how much the yield will be reduced in subclinical mastitis. Somatic cell counts have been linked to significant losses from both individual cows and herds; a higher count indicates a bigger loss.
- Milk with mastitis is also detrimental to human health. With severe clinical mastitis, gross milk abnormalities are immediately visible, and as a result, producers discard the milk. Such milk often wouldn't climb the food chain. However, if there is subclinical mastitic milk or if accidentally a little amount of badly contaminated mastitic milk was mixed with bulk milk, there may be changes in the milk. Certain mastitic milk bacteria can cause serious illnesses in humans. Even though all human pathogens are expected to be eliminated by pasteurization, drinking raw milk or performing subpar pasteurization raises alarm. Some strains of Staphylococcus aureus have the ability to produce heat-resistant enterotoxins, which can cause food poisoning that is harmful. Another matter of public health concern related to mastitis is antibiotic residues in milk. Food residues can cause severe reactions in those who are allergic to antibiotics, sensitize healthy persons at low amounts, and promote the formation of bacterial strains that are drug-resistant.

Producers must follow the suggested withholding time in order to lower the risk of antibiotic residues in milk.

Causes of mastitis

Poor management, ineffective milking methods, faulty milking equipment, inadequate housing, and breeding for an endlessly higher milk output are all potential causes of mastitis. Cows are exposed to more mastitis-causing microbes, their inherent resistance to the condition is decreased, or the teat canal is made easier for them to enter as a result of interactions between management and environmental factors. Climate, season, herd size, living conditions, diet, and stress all have an effect on the occurrence of mastitis. Physiological and genetic factors like pregnancy, milk flow, milk supply, and lactation stage also interact with these factors. The majority of dairy farms in the tropics do a poor job of controlling mastitis due to management and environmental variables. Smaller farms shouldn't be more challenging to control than larger ones. Cows everywhere in the globe depend on owners and herdsmen to keep management at a high level through careful observation.

Tests for mastitis detection

Many different cow-side and laboratory tests can be used to detect the presence of mastitis.

Physical examination

The most common test includes physically inspecting the udder. When the udder is getting ready for milking, this is done. There will be obvious heat and bigger quarters. It is advisable to do a thorough check after milking.

Strip test

The strip test, which includes pouring the first few streams of milk onto a strip plate or the ground in milking parlours, is the second most used test. Milk with mastitis will include clots, discoloration, or other abnormalities. As this practice will help the spread of mastitis germs to other cows or the other quarters, milk should never be drained straight into the hands or into the floor of a stanchion barn where the cows sleep at the same location they are milked. Dairy farmers should utilise the strip test before each milking to help diagnose clinical mastitis. The first milk is removed to accelerate milk let down and reduce the milking cycle.

Paddle test to detect somatic cells

Atypical milk can be found using the California Mastitis Test (CMT) and other tests of a similar nature. These tests essentially serve as management tools to help identify cows that may need extra care. Milk's somatic cells react with the test reagent to form a gel. The amount of gel that forms when milk is combined with the reagent determines how well a reaction performed. Before the test, a few streams of milk should be discarded, and the data should be noted for later use because gel formation is a very transient process. The test is useful in identifying subclinical infection that could otherwise go unnoticed when this is done.

Laboratory tests

For mastitis, particularly the subclinical variety, many inflammatory mediators can be evaluated, including somatic cells, serum proteins, enzymes such NAGase (N-acetyl-D-glucosaminidase), salt, and chloride concentrations. Comparing data from other materials, nevertheless, requires great caution because changes in inflammatory parameters don't always occur at the same time as the inflammatory process. Aseptically collecting milk samples from individual quarters or composite samples from all four quarters and sending them to a lab that specializes in mastitis microbiology for culture is the most precise technique to identify affected cows.

Management of mastitis

Dairy farmers in tropical regions frequently believed that mastitis wouldn't happen on their farms since they aren't used to paying much attention to the mammary glands. However, the disease was emphasized by the commercial dairy cattle industry. Mastitis management is made more challenging when some dairy farmers conceal the reality of the true incidence of mastitis in their herd out of embarrassment about their knowledge and management shortcomings. Inadequate milk testing at the bulk of the milk collecting centres' reception worsens the situation.

Because mastitis is usually unnoticeable and most dairy farmers are used to live with it, many of them seem reluctant to pursue mastitis control programmes. Tropical dairy farmers will need to be more aware of every aspect of teat exposure and resistance to microbes, both between milking and when the cows are being milked. When milking is done in the same location as the cows live, like in the stanchion barns, these factors include housing management, hygiene, nutrition, lactation stage, and methods used in milk production like pre-milking udder preparation, proper milking equipment and technique, and post-milking management. The advisory team or extension officers must be well-versed in the disease, grasp it correctly, and be aware of how to control it. Leaders in the dairy sector should also come to consensus on shared objectives and suggestions and make sure that everyone connected to the sector recommends the same things. Otherwise, farmers can grow perplexed and resist implementing new techniques.

Nutrition

Infection susceptibility is increased by malnutrition. It is challenging to meet the nutritional needs to handle the production when dairy cows are bred to produce more milk. There could be slight nutritional deficits. Mastitis is more likely to develop in cows with parturient paresis (milk fever), ketosis (energy deficiency), or deficiencies in selenium, vitamin E, and copper. The mammary immune system also needs vitamin A and beta-carotene to operate effectively.

Mastitis management approach

Mastitis is better understood as a herd issue than as a problem with a specific cow. Mastitis cannot conceivably be completely eliminated on a wide basis, unlike other cow diseases like brucellosis or tuberculosis. The unit of control is each distinct herd acting on its own. There is no connection between the severity of mastitis in one herd and that in neighboring herds. Large-scale mastitis control primarily requires a herd mastitis management programme.

A "control programme" is a series of actions that includes controlling mastitis. Such a programme needs to be practical, simple to understand, highly effective in the majority of dairy herds, increase economic returns, decrease new infections, decrease the length of time pre-existing infections last, provide verifiable proof that clinical mastitis is decreased, and be easily modifiable as better techniques are developed. In order to estimate the severity, origins, and risk of the infection spreading, the amount of infection in a herd must be recognized in order to control mastitis. The percentage of cows or quarters with the disease at any one moment is referred to as the prevalence of mastitis. Clinical or subclinical infection levels vary from farm to farm depending on the environment and management practices used in each herd. Each farm needs to be regularly monitored. Practically speaking, a mastitis control program's efficacy is determined by the amount of infection, or the percentage of infected cows or quarters, which depends on both the frequency of new infections and the length of each illness.

Best Recommended milking procedures for Mastitis management

- Even though the cow is typically only exposed to the act of milking and related activities for less than 15 minutes per day, this is when new diseases are most likely to develop. Therefore, it's crucial to follow the right milking processes whether cows are milked by hand in traditional dairying settings or by modern milking machines. Reducing the quantity of contaminating microorganisms on teat skin and promoting milk let down are two benefits of udder and teat preparation for milking. Correct teat and udder care will decrease microbial contamination of milk, the amount of milk left in the udder after milking, milk yield, milking time, and the spread of infectious and environmental organisms that may cause mastitis and/or affect milk quality. Here is a list of suggested milking techniques:
- The setting in which the cows are milked should be as hygienic and clean as is practicable. The cow should not be scared or agitated before milking because this causes hormones to be released into the bloodstream that interfere with normal milk let down and may lower the cow's natural resistance to illnesses like mastitis. The milking time should be a regular routine. Crossbred cows with hairy udders should have their long hair cut as often as necessary to remove it and lessen the quantity of bedding, manure, and dirt that could taint milk. Additionally, before milking, milkers should properly wash, dry, and rinse their hands with a disinfectant solution between cows. By following this approach, the transfer of mastitis bacteria from cow to cow will be minimized.
- A physical examination of the udder with the hand and an examination of the foremilk with a strip cup or plate before to each milking can both reveal the existence of clinical mastitis. These processes help identify milk that is clotty, stringy, or watery, as well as heated, hard, and swollen quarters. After each milking, strip cups and plates should be cleaned and sterilized to stop the spread of mastitis bacteria. Milk should never be drained into the hand directly since doing so transfers bacteria from cow to cow and teat to teat.
- The pituitary gland, located at the base of the brain, receives a signal from the proper washing and massage of the udder and releases the milk-letting hormone oxytocin into the bloodstream. The muscle cells around the milk-secreting tissues are then stimulated to contract by this hormone, resulting in milk let down. The lower udder and all of the teats need to be carefully cleaned, massaged, and dried. When keeping cows in stanchion barns, a sanitizing solution should be used in a bucket with

a specific cloth or paper towel to wash the udder. Avoid using a regular cloth or sponge because they can become extremely polluted and are nearly hard to sterilize.

- Whatever technique is utilized to clean the ventral surface of the udder and the teats, surfaces must be dried immediately after washing. Although individual cloths for each cow can be used if they are cleaned, sterilized, and dried in between milkings, single service paper towels are preferred.

Control of infectious mastitis

Although Corynebacterium bovis is also prevalent in many dairy herds, Staphylococcus aureus and Streptococcus agalactiae are the two most significant infectious pathogens. Although it is extremely uncommon, Mycoplasma bovis is nevertheless extraordinarily contagious. Although Staphylococcus aureus quickly colonises teat skin and the teat canal and has even been isolated from the udders of heifers that have never calved, diseased quarters are the main repository for these organisms. With the exception of Mycoplasma, these organisms are typically relatively readily controlled by maintaining good udder cleanliness, using good milking equipment correctly, dipping the teats after milking, and treating all of the cows' quarters before drying off. Staphylococcus aureus can be contained within a dairy herd by milking younger lactation animals first, older cows second, and known infected cows last. Rather than replacing the control programmes mentioned above, the use of vaccinations against this pathogen should be seen as an adjuvant. By using Comprehensive mastitis control approach in conjunction with a treat programme, Streptococcus agalactiae can be completely eliminated from a dairy herd in a matter of months. Before implementing specific methods for eliminating the microorganism, it is also essential that the situation be assessed by a knowledgeable individual.

Management of environmental mastitis

Mastitis can be brought on by a variety of environmental infections. Streptococcus uberis, Streptococcus dysgalactiae, and coliforms like Escherichia coli and Klebsiella spp. are among the most significant organisms in this group. Since these creatures are, as their name suggests, environmental in origin, the best way to control them is to limit their exposure to the teats. This entails maintaining a clean environment for cows. When muddy circumstances are present, during the wet season, and in housed herds as opposed to pastured herds, infection with these organisms is more common. More than inorganic bedding such as sand or limestone, organic bedding such as straw promotes the growth of environmental microbes. High rates of clinical mastitis in the herd, which can vary seasonally, are linked to environmental mastitis. In most clinical instances, Streptococcus uberis or gram-negative bacteria are identified. If E. coli is the predominant cause of the mastitis, sampling the entire herd may not be very helpful because E. coli mastitis infections rarely progress to chronic disease. A dry atmosphere is typically the solution to an E. coli mastitis problem. In the barn, water usage ought to be maintained to a minimum. Dry cow therapy and teat dipping have little effect on coliform mastitis.

Infections with Streptococcus uberis frequently happen during the dry season. The cow and its immediate surroundings are typically the origin of infection. Since antibiotic therapy is not always effective, persistent cases must be eliminated. The most important factor in preventing Streptococcus uberis mastitis is good hygiene. Animals with klebsiella infections frequently have to be culled because they are so difficult to treat. A few cows may have already died before Klebsiella was identified as the mastitis problem's cause. In some countries, a coliform mastitis vaccination is also offered. A heat-killed Escherichia coli J-5 mutant vaccine tested at the University of California, Davis can specifically stimulate antibody production against a variety of coliform bacteria, including Escherichia coli, Klebsiella pneumoniae, and Enterobacter species. It is given subcutaneously at drying off, 30 days later, and again within 14 days of calving. The vaccine won't be able to completely eradicate coliform illnesses from a herd, but when used in conjunction with sound management techniques, it will lower the likelihood of new infections.

Management of mastitis caused by other micro-organisms

The opportunistic microbes, which are frequently the most frequently isolated bacteria in the herd, comprise approximately 20 species of Staphylococcus other than S. aureus. These organisms typically cause only minor infections that don't result in much of a rise in somatic cell count. The organisms are typically found on the hands of milkers and on healthy teat skin, placing them in an advantageous position to colonise the teat canal and access the udder's milk-producing tissues. The percentage of infected quarters is high at calving because staphylococcal infections are more common during the early dry phase, when the teat skin is not exposed to germicides. The most effective method of preventing infections caused by Staphylococcus species is the use of a post-milking teat dip. At the end of lactation, dry cow therapy will successfully treat the majority of infections that were present; nevertheless, additional infections can be anticipated during the dry period.

Lactation therapy approach

Mainly Streptococcus agalactiae infection, a herd somatic cell count of 600,000, and clinical mastitis indicate the need for treatment during lactation. The kinds and drug sensitivity of the infecting microorganisms, as well as the intramammary circumstances, affect the treatment's success rates. It is advised that a consultant veterinarian create the treatment plan, basing their decision on previous herd experience, clinical symptoms, and environmental factors. The selection of medications should be based on the findings of sensitivity tests and bacteriological diagnosis.

Using the proper inflammatory indicators, such as changes in milk's somatic cell count, protein, enzyme, and electrolyte composition, one should be able to track how well mastitis treatment is working. The dairy farmers can also utilize the straightforward California Mastitis Test (CMT), which counts the somatic cells in milk. Although the test is somewhat subjective, when quarters are compared, its reliability is increased. It is important to be mindful of the possibility of infection spreading to other cows or quarters.

Acutely unwell cows may require the administration of the milk let down hormone oxytocin, significant amounts of electrolytes, anti-inflammatory prostaglandins products, and other medications as prescribed by the herd veterinarian. Frequent milking out of the udder can assist empty the udder of cellular waste and eliminate infectious organisms and their poisons, which can help the animal to recover.

Combination therapy, a procedure in which cows are treated with compatible drugs simultaneously in the udder and throughout the body, may result in a significantly higher concentration of antibiotics in udder tissues than treatment in the quarter alone, allowing drugs to reach micro-infections deep within the tissues. The likelihood of a cure for chronic cases can be doubled.

Dry cow therapy approach

At the end of lactation, dry cow therapy involves administering intramammal antibiotics to the udder. One of the most crucial elements of the "Comprehensive approach for mastitis control" is the long-acting antibiotic treatment of all quarters of all cows at drying off. It is the ideal moment to treat infections that are still asymptomatic. Dry cow therapy has several benefits, including a higher cure rate than when administered during lactation, the safe use of higher long-acting antibiotic concentrations, a decrease in the incidence of new infections during the dry period, the opportunity for damaged tissue to heal or regenerate prior to freshening, a decrease in clinical mastitis at freshening, and the avoidance of drug residues in milk. A quarter that becomes sick during the dry season or that is still infected from the previous lactation would yield 30 to 40% less milk than a quarter that has been treated for infection at drying off and cured at calving. No indication of treatment-associated resistance to medications among major or minor mastitis pathogens, allaying worries that routine dry cow therapy would increase the resistance of mastitis micro-organisms to routinely used drugs.

Bringing the cow to the end of lactation and treating the animal after the last milking, are the suggested steps for treating dry cows. Milk will fill the udders up until a specific hydrostatic pressure is reached before being reabsorbed into the bloodstream. The beneficial vehicle for delivering medications to all sick locations throughout the udder is milk in the udder during drying off. Therefore, the cow shouldn't be fully dry when receiving treatments.

Treatment method

Teat ends must be carefully cleaned by swabbing for a few seconds with with 70% alcohol before administering any kind of therapy into a quarter. Another practice that many vets employ is to first dip the teat in a teat dip before administering the therapy through the dip. Microorganisms that are on the end of the teat and in the teat canal may be driven into the quarter and produce more severe mastitis than the one for which the antibiotic was designed if teats are not cleaned before treatment. The technique used to inject the medications into the udder is also very significant. When a cannula is pushed all the way through the teat canal, for instance, it may actually push microorganisms into the quarter. The cure rate is increased and the risk of new infection is decreased by inserting the infusion cannula about three millimetres into the teat canal. Infusion cannulas have now been created by the majority of pharmaceutical companies to these criteria. To help prevent their milk from getting contaminated with milk from untreated animals, treated animals should be properly labeled in an evident way and kept apart from untreated animals.

It is essential that all label instructions be followed in order to prevent detectable residues in milk. It is important to pay close attention to dosage amounts, delivery methods, and withdrawal intervals. Aside from intramammary infusion, other methods of administering antibiotics that result in lengthier withdrawal times include intramuscular injections, uterine boluses, and antibiotic feed. Cows that have received treatment ought to be kept apart and prominently marked. Additionally, all treated cows should have written records retained, since this data can be used to guide culling decisions for chronic cases.

A herd with a subclinical mastitis problem may have a high somatic cell count, a high bacterium count, a high rate of clinical mastitis, or a combination of the three. Mastitis problem herds typically have one or more of the following contributing factors: unsanitary milking practices, unsanitary surroundings, broken milking equipment, inappropriate milking techniques, and insufficient treatment options. The best place to start when dealing with a problematic herd is by doing the now commonly used recommended method on a sample of herd milk in a lab. The process gives accurate and timely information regarding the cleanliness and udder health of a particular herd's milk production. To determine the types and quantities of various microorganisms in the sample, the process entails three tests: a somatic cell count, a bacterial count (standard plate count), and microbiological culture on blood agar. A veterinarian or skilled technician can then assess the combined results of the three tests to determine the nature, severity, and most likely cause of a herd mastitis issue. They can also spot a herd hygiene issue as shown by improperly cleaned milking equipment, wet milking, contaminated water, or inadequate milk cooling. The data essentially serves as a road map for choosing the appropriate course of action.

Mastitis will be common in any herd with a high somatic cell count (500,000–1,000,000), which should be regarded as a problem herd. Each quarter should have a few streams of foremilk that are thoroughly inspected for anomalies. Cows whose milk is obviously abnormal or tests positive for the CMT need to be separated, and the milk should not be added to the bulk tank. To find the microorganism causing the illnesses, a somatic cell count and laboratory culture of the quarter milk should also be performed. This information is important for deciding whether to treat and it may help identify the cause of the issues and the necessary corrective measures.

A high bacterial count suggests the possible presence of mastitis-causing microorganisms and/or bacterial contamination brought on by improper milking technique, dirty milking equipment, and/or inadequate milk cooling. If streptococcal organisms make up more than 75% of the total, infected udders are most likely the source. If the streptococcal count is less than 25% of the total, wet milking, inappropriate udder preparation, or inadequate milk cooling are likely to be the culprits. Large amounts of staphylococci, coliforms, spore formers, and other microorganisms in samples along with large numbers of streptococcal bacteria frequently point to both diseased cows and improper udder preparation as problems (wet milking).

A sample with a high staphylococci count typically suggests that the milk wasn't properly cooled. Rarely, if ever, are high staphylococcal numbers in herd milk the result of a single infected udder. High coliform counts frequently point to damaged teat cup liners, cold water used while washing the milking system, milkstone on milk-contact surfaces, and improper cleaning and sanitizing agents used on milking equipment. Large concentrations of coliforms, staphylococci, and environmental streptococci in a sample are frequently signs of improper milk cooling. The finest sets of instructions for mastitis management are, above all, the Comprehensive plan of mastitis control supplemented with the Recommended Milking Procedures described previously.

Mastitis management in heifers

Even in relatively young animals, heifer udder infections are quite common and can lead to tissue damage and inflammation before calving. Environment, insects, particularly flies, suckling among calves (especially those fed mastitic milk), and microflora in the mouth and on the skin and hair coat are thought to be sources of infectious microorganisms. Mastitis in heifers can be avoided by utilizing management techniques such fly control, individual calf pens, and separating pregnant heifers from dry cows. Early-lactation mastitis is decreased by strategically administering antibiotics to pregnant heifers at various times before the first parturition.

Heifers that are known to have Staphylococcus aureus-related udder infections should be given a dry cow treatment product eight to ten weeks before giving birth. The ideal method for treating other illnesses is to use a nursing cow product seven to ten days before to the anticipated calving date. About 90% of the infections will be wiped out by this practice. Heifers that have received treatment will produce 10% more milk than untreated heifers. When treating heifers, extreme care must be taken to ensure that the teat ends are well cleaned before treatment and that the teat cannula is only partially inserted.

A team of veterinarians, laboratory microbiologists, milk plant field representatives or extension officers, milking equipment dealers, and most crucially, dairy farmers should carry out an efficient mastitis control programme. Extension agents frequently send milk samples to the lab for the diagnosis of mastitis. They are primarily concerned with environmental issues and milking methods. Dealers of milking equipment should make sure the unit is properly maintained, cleaned, and functional. Veterinarians that are currently in practice should develop the programme. The dairy farmers who must be enthusiastic and highly motivated to ensure the success of the initiative should receive advice and education.

Foot-and-mouth disease

One of the most significant economic disease known to man for a very long time has been foot and mouth disease (FMD). Foot and Mouth disease affects animals with cloven hooves, including cattle, buffaloes, sheep, goats, and pigs, among others. It is a highly contagious viral vesicular disease. High temperature (104–106 degrees Celsius), lack of appetite, lethargy, excessive salivation, vesicles in the mouth, particularly on the gums and tongue, which cause ulcers, ulcers and wounds in the hoof in the inter-digital area, blisters on teats, etc. are clinical and visual indications. FMD causes a drop in milk production, a slowdown in growth, infertility, a reduction in the working capacity of bullocks, and an international trade ban. According to economic effect, it is one of the most significant animal illnesses, and it is widely acknowledged as a priority for control and elimination. Farmers experience significant financial losses as a result of this illness, which last throughout the life cycle of the animal. Despite India being the world's largest producer of milk, FMD also act as hurdle for access to export markets (Estimated national loss due to FMD – Rs.20,000 crore per annum – source ICAR).

Close contact with diseased animals, tainted food and water, animal movement, aerosols, and contaminated objects are all ways FMD spreads. Once the animal has become infected, there is no rapid cure. The infected animal needs to be kept apart, its symptoms need to be treated, and the animal shed needs to be cleaned with the right disinfectant. By routinely immunizing large numbers of susceptible livestock at regular intervals until the disease's incidence declines, FMD can be controlled. This will open the path for the disease to be eliminated gradually from the nation. But for two primary reasons, this significance is unquestionably increasing on a worldwide scale. Firstly, domestic cattle animals are more productive today than they were 50 years ago, making them more vulnerable to FMD's effects in many parts of the world. Second, the likelihood of virus propagation grows as there is a greater worldwide exchange of people and cattle products.

Geographically, by production methods, by sectors, or by socioeconomic level, this growing global influence is not allocated fairly. Complicated infrastructure and fencing required to keep cattle and buffalo apart, preventing FMD in wide areas. Despite the nearly universal infectiousness of the FMD virus, decisions about where, when, and how the illness is controlled, who should bear the costs, and who would benefit most, vary greatly as a result of these diverse affects. And economics plays such a significant part in aiding such judgments. India's domestic cattle population is affected by the foot-and-mouth disease (FMD), which costs animal owners a lot of money. The disease's clinical manifestation is easily seen in vulnerable livestock, primarily cattle, buffalo, and pigs, but it remains prominent in carriers. In order to restrict the spread of the disease, India has implemented a number of procedures, starting with field-based clinical diagnosis of the illness, sending clinical samples for laboratory analysis, and choosing potential vaccine candidates. To protect any susceptible livestock population, vaccines are utilized. This should reduce any financial losses caused by the disease to animal owners. Since 2003–2004, the Government of India has been conducting an intensive FMDCP in a phase-by-phase fashion, and by 2017–2018, it has reached all of the nation's districts. The FMDCP plans to vaccinate all of the country's susceptible livestock, including cattle, buffalo, sheep, goats, and pigs. The goal of that activity was to eradicate the disease from the nation by 2025–2030. The South Asian Association for Regional Cooperation (SAARC) region is being served by the Directorate on FMD, which is working nonstop in this regard.

Brucellosis

Due to its zoonotic potential, brucellosis is a livestock reproductive disease that can cause significant financial losses and harm to human health. This disease can constantly be contracted and spread among farm workers and livestock owners. Therefore, eliminating Brucellosis will benefit both human and animal health, as well as result in significant financial rewards for farmers and owners of animals. The bacteria Brucella abortus is the source of the reproductive illness brucellosis in cattle and buffaloes. Fever, sterility, delayed heat, interrupted breastfeeding leading to the death of calves, and decreased production of milk are some of the symptoms of the illness. It also triggers abortion during the latter stage of pregnancy. Due to its zoonotic nature, brucellosis has a negative influence on human health. Additionally, it poses a major occupational risk. Humans who are afflicted may experience a fluctuating fever, nocturnal sweats, body aches, weakness, and poor appetite. Bovine brucellosis is an endemic disease in India and has recently appeared to be spreading more widely, possibly as a result of greater trade and quick movement of livestock. Due to intimate contact with animals, the nature of livestock farming, particularly in rural India, is conducive to the spread of illness from livestock to humans. The disease can be avoided by immunization in the absence of a treatment for Brucellosis in bovine animals. A once-in-a-lifetime vaccine of female bovine calves can control Brucellosis (4–8 months old). Therefore, it is crucial to prevent Brucellosis by immunizing all female bovine calves (4 to 8 months old) and Foot and Mouth Disease (FMD) by immunizing all cattle, buffaloes, and goats in the nation. In addition to improving animal health, this will increase animal productivity and the acceptance of our animal products around the world. Finally, efforts in this direction would help double the revenue of farmers. The production of dairy cows is influenced by programmes for disease control, herd management, and good health care. Similar to health, nutrition is crucial and should be seen as a key element of herd management. Farmers must identify sick animals as soon as possible, isolate them from the rest of the herd, and start treating them. Early diagnosis and treatment will be a beneficial management tool if you keep an eye out for any potential issues within the herd. A veterinarian should make additional diagnoses if an animal stops eating, stops producing milk, coughs, has diarrhoea, is lame, or has an odd discharge. To prevent the transmission of contagious diseases, sick cows should be kept separate from healthy ones. In addition to individual animals, the entire herd should be examined and diagnosed.

Good feeding, breeding, care, and management, together with good record keeping and a dairy health programme, are all important components of running a dairy farm profitably. The goals of the health programme are to stop the spread of infectious illnesses and parasites as well as the introduction or occurrence of sickness. Government regulations that aim to eliminate epidemic diseases in dairy cattle typically depend on local expertise. Government and international organizations are accountable for the diseases that affect humans and are spread by dairy cattle and dairy products. Dairy cooperatives must take health care into account, especially diseases that might spread from cattle to people and have an impact on milk spoilage. In addition to offering efficient services, veterinary services and dairy extension professionals should place a strong emphasis on continuing education for farmers in the areas of farm management and disease control. However, in the end, it is the smallholder who needs to comprehend the benefits of excellent management techniques and the connections between dietary intake, housing quality, and health.

9

Feed and Fodder

The proper supply of feed and fodder is a requirement for the growth and development of the animals. Over the past few decades, there has been a considerable growth in milk production, with India now producing the most milk in the world (187.7 million tones in 2018–19). The main causes of India's low milk yield are extrinsic (poor nutrition/feed management) as well as intrinsic (low genetic potential). According to empirical research, improving the quality and quantity of feed and fodder has a bigger impact on raising milk yield than improving the breed. A key factor impacting the production of farm animals in the nation is the lack of timely availability of nutrient-rich feed and fodder. Although the availability of animal feed has improved throughout time as a result of increased agricultural production, the supply has never been sufficient to meet overall demand. The shortage of dry and green fodder in India in 2020 was estimated by the Indian Grassland and Fodder Research Institute (IGFRI), Jhansi, to be about 12% and 30%, respectively. The productivity and production of cattle are negatively impacted by this nutritional shortfall. In India's livestock industry, low livestock productivity is a big issue. One of the biggest problems in India is raising the output of agricultural animals. Low livestock productivity means that there aren't enough foods from animals in human diets. The scarcity of green fodder and high-quality feed severely limits the output of dairy animals in many developing nations, including India. The primary sources of forages for feeding animals in these nations are dry grasses and crop wastes. These crop leftovers are high in fibre and lignin but poor in nitrogen. Animal intake and digestion are limited as a result. If nutrients are added, the nutritional value of these crop wastes can be improved. According to estimations from the ICAR-NIANP, the gap between the amounts of dry feed, green feed, and concentrates needed and actually available in 2015 was 21%, 26%, and 34%, respectively. This is likely to increase to 23 per cent, 40 per cent, and 38 per cent, respectively, by 2025. The fodder deficit in India in terms of green fodder, dry fodder, and concentrates was 26 million tones (MT), 21 MT, and 34 MT in 2015, which is expected to reach 40 MT, 21 MT and 38 MT by 2025 respectively.

Table 9: Feed and fodder estimates in India

Feed	Particulars	Year		
		2015	2020	2025
Dry	Requirement	491	530	550
	Availability	387	408	433
	Deficit (%)	21	23	21
Greens	Requirement	840	880	1000
	Availability	619	596	600
	Deficit (%)	26	32	40
Concentrates	Requirement	87	96	105
	Availability	58	61	65
	Deficit (%)	34	36	38

This scarcity results from inadequate attention being paid to the cultivation of fodder crops and increased pressure on land for growing food grains, oilseeds, and pulses. Due to the pressure on the land for agricultural and non-agricultural activities, the grazing pastures are gradually getting smaller while the livestock is expanding quickly. The majority of grazing pastures have either been encroached upon or deteriorated, reducing their availability for grazing. About 4% of the agricultural area is devoted to fodder farming, and this percentage has been constant for the past forty years. The need for both green and dry feed is anticipated to rise significantly to 1,012 million tones and 631 million tones, respectively, by 2050 due to an expanding livestock population and governments focus on genetic improvement of cattle through cross-breeding programmes. After 2012, the wholesale price index for cattle feed increased far more quickly than the index for milk. Concerns have been observed about the ingredients' high price volatility. Therefore, businesses frequently substitute feed ingredients at the expense of feed quality, the most crucial factor that has a wide range of effects on both human and livestock health as well as animal productivity. This is done in an effort to maintain profit margins.

Strategies for future

The concept of using land for numerous crops in a sustainable way must be adopted. The availability of pasture for the animals might be significantly increased by using silvi-pastoral and horti-pastoral models appropriate to the region. In the nation, open forests with less than 0.4 canopy density make up about 29 million hectares of land that can be planted with fodder trees. This vast amount of space can be used to grow fodder instead of only an understory on partially shaded ground without harming existing trees. The non-cropped parts of agricultural land that are not farmed, such as bunds, pond embankments, basins of plantation and horticulture crops, hedges with fodder crops, etc., should also be highlighted.

Farmers may benefit from the promotion of the growth of many types of green fodder, including napier, marvel grass, moringa, maize, bajra, jowar, cow peas, velvet beans, thorn-less cactus, oats, berseem, rye grass, and Chinese cabbage. Initiatives include establishing fodder cooperatives and expanding the amount of community grazing lands to ensure that there is a sufficient supply of feed and fodder for the milk cows to support the production of milk. To ensure a sufficient water supply for the production of fodder, water irrigation cooperatives could be connected to the cooperatives for fodder. In order to increase animal production, a number of State research organizations and the IGFRI have created enhanced fodder crop varieties and technologies that can guarantee year-round availability of high-quality feed and fodder.

To protect the interests of small dairy producers, it is essential to establish a community fodder bank where surplus fodder can be kept as hay, silage, or fodder blocks for use during times of scarcity. Additionally, focusing on the following four tactics will be beneficial: The silage business model involves seed companies, operational service providers (for baling and supply chain services), and rural retail channels. It also involves improving the extension network and smoothing out the credit facility for fodder production. These will eventually open the door for the Indian dairy herd to reach its full potential.

Additionally, there are regional feed disparities in India, with certain regions having an abundance of straw and green fodder while missing these in other areas. Millions of dollars' worth of straws are burned in the fields of north Indian states. In addition to wasting a feed supply, this also pollutes the ecosystem and degrades the soil. The management of feed resources in India has to be improved. Using specialized balers to harvest straw from fields and then processing it to create commercially viable, well-balanced animal diets would be one improvement. If pushed among livestock farmers, feed supplementation technology such urea molasses mineral blocks can efficiently address mineral and other nutritional deficits.

In addition to making optimum use of crop leftovers, it's critical to focus on agro-industrial waste products and unconventional feed sources, such as cactus and tree leaves, to increase the supply of feed resources. There are currently a lot of technologies and methods for enhancing livestock feeding in India, but they still need to be implemented at the farmer level. Sustainable development depends on making effective use of the present feed resources, which includes reducing waste and expanding the base of feed resources by investigating new feed sources, especially those that don't compete with human food. Similar to many other nations, India produces a significant amount of fruit and vegetable waste from the processing, packing, distribution, and consumption of fruits and vegetables.

In addition to making optimum use of crop leftovers, it's critical to focus on agro-industrial waste products and unconventional feed sources, such as cactus and tree leaves, to increase the supply of feed resources. There are currently a lot of technologies and methods for enhancing livestock feeding in India, but they still need to be implemented at the farmer level. Sustainable development depends on making effective use of the present feed resources, which includes reducing waste and expanding the base of feed resources by investigating new feed sources, especially those that don't compete with human food. Similar to many other nations, India produces a significant amount of fruit and vegetable waste from the processing, packing, distribution, and consumption of fruits and vegetables. To make the most use of the feed and fodder resources available, animal feeding strategies such the use of a compressed complete feed block, silage, fodder chaffing, and soaking should be employed. Campaigns must be started by extension and advisory agencies to encourage farmers to feed their cattle keeping new options in mind.

10

Importance of Mineral Mixture for Dairy Animal

In terms of their contribution to human nutrition, the dairy industry is very important. Since dairy farming has the ability to offer meaningful self-employment and stable income, it is a dependable source of subsistence for the rural poor. Additionally, it requires less labour and cash to maintain. Dairy cattle feeding directly affect the animals' ability to reproduce and produce. Therefore, the addition of a mineral mixture is necessary to maximize the animal's productivity and milk output. For optimum milk production, reproductive success, and herd health, dairy cattle need at least 17 minerals and three vitamins in their diet. Despite the rarity of conventional mineral or vitamin deficiency symptoms, under- and overfeeding of several minerals and vitamins does happen frequently. Even little deficits or imbalances have the potential to cause issues with milk production, health, and reproduction.

Poor growth, deficient illness, decreased milk production, reproductive issues, and impaired immunity are all symptoms of mineral insufficiency in farm animals. Significant progress has been made over the past ten years in our understanding of how trace-element supplements affect dairy cows' ability to produce milk. Because several of them, including Mg, S, K, I, Co, Fe, and Mn, are available in feeds and fodders of particular regions at the required quantities, supplementing the diet with all the elements may not always be desired. Deficits in minerals including Ca, Cu, and Zn have been linked to significant alterations in key physiological and biochemical markers.

One of the main issues with livestock production in India is low reproductive efficiency in cattle and buffaloes, which leads to poor economic returns for livestock farmers. One of the key causes of the low reproductive efficiency is the lack and/or imbalance of micro and macronutrients. Infertility affects 31% of all cattle populations in India. In the majority of India, the practice of giving animals extra minerals is uncommon. Wherever it is used, the mineral mixture is added to the compound feed without considering the local mineral conditions, which results in mineral imbalances. Supplementing the minerals that are most lacking in that region could be more practical and cost-effective because it eliminates the negative consequences of having too much of another mineral.

Functions of different minerals

Calcium: Essential for milk production, bone and teeth formation, contraction of muscles

Phosphorus: Milk production and energy metabolism

Magnesium: Important for the integrity of bone and teeth and involved in protein synthesis and metabolism of carbohydrates and lipids

Sulphur: Required for protein synthesis and metabolism of carbohydrates and lipids, and Sulphur is a part of B-complex vitamins, thiamin and biotin

Sodium and potassium: It is required for maintenance of osmotic balance and required in acid- base equilibrium

Copper: Required for haemoglobin synthesis, tissue pigmentation and component of several metallo-enzymes and required for normal reproductive functions

Zinc: Required for normal functioning of epithelial tissue, spermatogenesis and the development of primary and secondary sex organs and it activates vitamin A and its deficiency leads to night blindness

Manganese: Activator in the synthesis of fatty acids and co-factor for many enzymes involved in carbohydrate metabolism

Iodine: Required for the synthesis of thyroid hormone and necessary for reproduction and growth of animals

Cobalt: Required for the synthesis of vitamin B by the rumen microbes, essential for haemoglobin synthesis, required for the metabolism of carbohydrates, protein and fat and helps in improving immune status of animals

Mineral Feeding

Forages and grains, which are natural feeds, and mineral supplements, which are used to balance the minerals in forages and grains, are the two main sources of minerals. Calcium, phosphorus, magnesium, potassium, sodium, chlorine, and sulphur are the main minerals (macrominerals) needed by dairy cows. Iodine, iron, cobalt, copper, manganese, zinc, and selenium are among the minerals that must be consumed in much smaller, trace amounts (microminerals). To attain peak performance and herd health, the right amount of a mineral must be provided, regardless of how much is required.

Macrominerals

Calcium and Phosphorus: In the mature dairy cow's calcium and phosphorus needs are influenced by her body weight, her milk output and composition, and her stage of pregnancy. The total lactation cycle requires a balance of calcium and phosphorus needs. To preserve cow health and performance, minerals that were mobilized during early lactation must be supplied before the next calving. To lessen the likelihood of milk fever, it is crucial to keep an eye on the calcium and phosphorus levels during the final four weeks of the dry period. For the dry season, the requirement for calcium is 0.39 percent of the dry matter of the ration, while the requirement for phosphorus is 0.24 percent. Dietary calcium levels of 0.80% and phosphorus levels of 0.50 percent of dietary dry matter are required for high-yielding cows. For maximal production, the phosphorus requirements specified for dairy cattle include a margin of safety. Because phosphorus is the most expensive ingredient in the majority of dairy diets, avoid huge dietary excesses. During lactation as well as the dry period, keep the calcium to phosphorus ratio between 1.4:1 and 2.5:1 for optimum health and performance.

Magnesium: For calves under three months old, the need is 0.10 percent; for high-yielding cows, it is 0.25–0.30%. In the beginning of lactation, when extra fat is fed, or when grass tetany conditions exist, higher magnesium levels may be necessary. For alfalfa-based rations, magnesium oxide is frequently employed as a buffer and resulting in ration magnesium concentrations of 0.30% or higher. According to research, feeding heifers 18 g of magnesium daily in the form of magnesium oxide five to six weeks before to calving helps minimize udder edoema. Magnesium-deficient forages are produced by soils that are low in magnesium or that are limed using calcium sources other than dolomitic limestone. Supplements such as magnesium oxide (56% magnesium) or dynamate (12% magnesium) should be fed when soil levels are low.

Potassium: Legume forages are a good source of potassium. Potassium levels may be on the verge of deficiency when brewers grains or corn silage are fed in large quantities. When heat stress is an issue, potassium needs are raised to 1.3–1.5 percent of the dry matter of the ration.

Sulfur: Again, legumes are a rich source of sulphur, as do protein feeds. Nitrogen to sulphur ratio of roughly 10:1 is ideal. Additional sulphur typically required only when the ration comprises of urea, maize silage, or hay of poor quality. Calcium sulphate (19% sulphur), sodium sulphate (10% sulphur), and methionine hydroxy analogue are all excellent sources of sulphur. Since trace mineralized salt contains a sizable amount of sulphur, it is not advisable to increase the diet's sulphur intake unless a specific shortage is known. Excessive sulphur hinders copper usage and raises the risk of molybdenum poisoning.

Sodium chloride: Dairy cattle's daily salt needs are easily satisfied by adding 1% salt to a grain mixture and providing more salt at their choice. Cows that are lactating require 50–100 g of salt per day. Dry cows require 40 g of salt per day. Reduce salt consumption during the final two weeks before calving if udder edoema is an issue.

Microminerals

In general, dairy cattle's needs for these trace minerals can be satisfied by adding trace mineralized salt to the grain mix, free choice trace mineral salt, and microminerals contained in the feeds.

Cobalt: The majority dairy rations will require no supplemental sources of cobalt.

Copper: The majority dairy rations need to be supplemented with copper either from trace mineralized salt or a premix containing a copper source such as copper sulfate.

Iodine: Twelve milligrammes of iodine should be consumed each day from trace mineralized salt or another iodine source. Do not give more than 50 milligrammes of iodine per kilogramme of body weight as this will increase milk iodine levels and result in toxicity issues such excessive nasal discharge and watery eyes.

Iron: After two months of age, iron deficiencies in dairy cattle are uncommon. A usual dairy ration contains much more iron than is actually needed by the cow.

Manganese: The majority forages, grains and protein supplements are simply fair sources of manganese so use supplements, particularly with high producing cows.

Zinc: The usage of other trace minerals, such as copper and iron, is hampered by excessive zinc levels. In most cases, feeding zinc over the suggested 40–60 ppm in the dry matter of the ration is unnecessary. Good sources include zinc sulphate and zinc oxide. Though additional research is required, zinc methionine may enhance cows' hoof health and milk output.

Selenium: When soils are weak in selenium, dry cows should have a supplement of three to five milligrammes per day, and lactating cows should receive six to eight milligrammes per day. Selenium supplementation may increase reproductive effectiveness and decrease retained placentas in dairy cattle. Selenium insufficiency has been linked to white muscle illness in calf.

Along with correct milking management and environment, adequate dietary selenium may also help to reduce the incidence of mastitis. Selenium should be handled cautiously, though, as it is dangerous if used in excess and is present in high concentrations in some soils. A decrease of appetite, tail hair loss, hoof sloughing, and even mortality occur at levels higher than 2 ppm.

Use of Chelated Minerals: In several research trials, using mineral chelates—or organic mineral complexes—instead of the same mineral's inorganic form improved performance and reduced somatic cell counts in milk. The effects of zinc, copper, manganese, iodine, and selenium chelates on milk production have been inconsistent, and it is unclear how useful they will be in practice. Further study is required to better identify the management situations where favourable production or health responses are anticipated before any firm suggestions can be made.

Mineral Feeding Methods

Force Feeding: With this approach, issues with palatability, daily and cow-to-cow fluctuation in intake, and excessive mineral consumption are all eliminated. Based on the mineral content of the forage and grain and the amount of milk produced, the producer decides how much of various mineral supplements to add. A correctly planned and mixed ration is essential. In a whole mixed ration, force feeding is most effective. The employment of a grain carrier is another often practiced force feeding technique. Because forage intake declines as milk production rises and grain intake rises, do not use moist forage alone as a mineral carrier.

Free Choice: Only trace amounts of mineralized salt should typically be provided to animals at their discretion because this method is less precise than force feeding. Cows might prefer a supplement owing to flavour, odour, or particle size, acquired appetite, or just pure appetite.

Topdressing: In stall or stanchion barns, this technique is frequently employed. In addition to their daily meal, cows are given 1 to 3 ounces of a trace mineralized salt or mineral-vitamin premix and 1 to 6 ounces of a calcium-phosphorus supplement. As a result, extra nutrients can be supplied before early lactation, when the body consumes its maximum amount of dry matter.

Fig. 19: Urea molasses mineral block preparation during experiential learning (*Source*: DUVASU, Mathura website)

11

Feeding Bypass Protein to Dairy Animals

For the purpose of passing into the lower gastrointestinal tract and supplying the cow with vital amino acids, dairy bypass protein (UDP – undegraded protein) is a type of animal or plant-based protein that resists breakdown in the rumen. More metabolizable protein is made available by avoiding the rumen, assisting the cow in meeting her needs for amino acids for milk production and its constituent parts. Bypass protein is used to adjust the dairy ration for maximum milk production and feed efficiency. Dairy farmers can minimize the amount of crude protein in the diet and enhance the amount of metabolizable protein by feeding high-quality bypass protein to their animals. This reduces ammonia production and nitrogen excretion.

The amino acids supplies for ruminant animals come from food protein sources that are resistant to rumen degradation and from microbial protein that is produced in the rumen. The solubility of the protein and the speed at which it enters the small intestine determine how much of it escapes rumen breakdown. The amount of plant organic matter fermented, ammonia levels, and mineral concentration in the rumen all influence how much microbial protein is synthesized. It is very obvious that animals' needs for amino acids are not entirely satisfied by the typical dietary protein sources. The researchers created the concept of protein protection from ruminal degradation with the primary goal of improving the supply of essential amino acids to the productive animal and reducing nitrogen losses as urea in the urine. Rapid and extensive degradation of valuable proteins in the rumen led to this concept's development. The cost of feed is raised by the more expensive price of protein supplements. We can either minimize the amount of protein in the diet or increase the productivity of the animals by optimizing the usage of protein supplements within the ruminant system.

Food protein is hydrolyzed by rumen bacteria into peptides and amino acids. The majority of amino acids, however, broke down quickly into organic acids, ammonia, and carbon dioxide. The main nitrogenous nutrient for bacterial growth is the ammonia that is generated. Some ruminal bacterial species directly synthesize microbial protein from peptides. The amount of dietary protein converted into microbial protein must be a significant factor in nitrogen economy because this protein creation is an energy intensive mechanism, hence care should be taken to reduce ruminal breakdown via artificial means. 25–50% of the nitrogen used by the rumen microbial community comes from sources other than ammonia. These peptides or amino acids, which are assumed intact, come from either the protein in food, recycled nitrogen entering the rumen, or the breakdown of bacterial and protozoal protein inside the rumen. In the rumen, free amino acids are an intermediate result of the breakdown of proteins by microbes. The two sources of UDP and microbial protein provide the host animal with all the necessary amino acids. Although the microbial protein created in the rumen is sufficient in the case of poor yielders, in animals with rapid rates of growth and production, the microbial supply is constrained by the need for amino acids, necessitating the provision of proteins in the form of UDP, escape proteins, or shielded proteins to meet the demand.

Protection of Protein From Ruminal Degradation

Several techniques, including heat treatment, chemical treatment or modification, inhibition of proteolysis activity, and discovery of naturally protected proteins, can be used to protect proteins. These methods improve the supply of amino acids without increasing ammonia production, which increases the animal's performance in compared to the traditional sources of dietary proteins. There are many ways to achieve protection, including:

1. Metal Amino Acid Complex
2. Heat Treatment
3. Formaldehyde Treatment
4. Oesophageal Groove closure
5. Post Rumen Infusion (Fistula)
6. Encapsulation of Proteins
7. Amino Acids Analogs
8. Coating of Protein with Insoluble Substances
9. Feed Processing Normal procedure
10. Lowering Ruminal Protease Activity

Effects of Bypass Proteins on Animal Performance

1. Extra supply of amino acids at intestinal and tissue level
2. Lesser ammonia production in the rumen because proteins are fermented to ammonia and low degradation of protein will lower ammonia level.
3. Lesser urea production in liver as ammonia is being absorbed at lower level which saves energy.
4. Surplus amino acids go for gluconeogenesis
5. Best exploitation of protein Resources

Beneficial Effects of Bypass Proteins

Growth Performance: Animal growth is significantly accelerated when bypass protein meals are given to them. Feeding bypass protein to calves and kids, significantly increase their growth rates. These animals' growth rates increased between 30 and 40%, it was discovered. Bypass protein is fed to growing animals to boost feed conversion efficiency in addition to growth rate. In reality, it not only lowers the cost of rearing but also causes male or female animals to reach maturity earlier due to the greater growth rate. In terms of these animals' improved reproductive efficiency, this is unquestionably a larger surplus.

Lactating Performance: The majority of feeding trails significantly increased the yield of milk. The range of the milk output increase was between 8% and 10%. Studies on animals with medium milk production levels showed that animals that produce 8 to 10 litres of milk per day can benefit from bypass protein feeding.

Reproductive efficiency: The juvenile stock can mature quickly and begin reproduction at a younger age due to the fast growth rate brought on by protein feeding. It has been demonstrated that feeding bypass proteins can increase the reproductive efficiency of breeding buffalo bulls and cross-bred bucks, both in terms of sexual behaviour, such as libido score, as well as the seminal characteristics, such as ejaculate volume, mass activity, and sperm count per ml. Similarly positive results were obtained in females, where the conception rate decreased after feeding bypass proteins.

NDDB has standardized and commercialized bypass protein technology, using locally available protein meals such as rapeseed meal, sunflower meal, groundnut meal, guar meal and soybean meal. Treated protein meal can be either fed directly to animals as top feed at the rate of one kg per animal per day or can be incorporated in cattle feed at the rate of 25 per cent and this cattle feed can be fed at the rate of 4-5 kg per animal per day, depending up on the level of milk production.

Supplementation of bypass protein to medium and high yielding growing animals increases the live weight gain, DM intake and milk production in animals.

12

Feeding Bypass Fat to Dairy Animals

Dietary fat, which is not degraded in upper part of digestive tract (rumen) of animal, but gets digested in lower alimentary tract is known as bypass fat. Rumen protected fat is another name for this bypass fat. Immediately following calving, dairy animals lose a significant amount of energy through milk, although their feed intake is decreased. Animals continue to have negative energy balances as a result. These facts have a negative impact on the animals' physiological conditions, reproductive abilities, and milk supply. The negative impact is greater in animals that produce more. Bypass fat is a rich source of energy. Additionally, bypass fat is an excellent source of calcium. Bypass fat supplementation therefore enhances animal physical condition, reproductive efficiency, and milk production. Animals who consume oil have stomach issues. Animals' upper digestive tract (rumen), where oil is digested, has lower energy availability than bypass fat. Oil cannot be used in place of fat to obtain extra energy because of this.

In India, buffaloes produce a large portion of the country's high-fat milk, and during the early lactation period, the amount of energy needed to maintain body tissues and produce milk frequently exceeds the amount of energy provided by the diet. This creates a negative energy balance, which forces the mobilization of body fat reserves to meet energy needs. The negative energy balance period frequently begins before calving because of the decreased feed intake towards the end of gestation. Beyond influencing peak milk yield and overall lactation yield, the negative energy balance during early lactation results in delayed post-partum ovarian activity. High yielding crossbred animals should be fed either more cereal grains or a diet supplemented with bypass nutrients during the first lactation to meet their energy requirements. However, consuming too much cereal grains can lead to rumen acidosis, which makes animals sick and causes them to stop eating, which reduces milk production. Bypass fat addition helps nursing animals produce more milk and fat.

Use of dietary fat in dairy animals

In the diet of early nursing or high yielding dairy animals, fat aids in boosting the food's 2.25 times higher energy density than carbohydrate. The addition of fat to animal diets prevents laminitis, acidosis, and negative energy balance during early lactation. It also controls heat output and incorporates fatty acids into milk fat. If raw edible oil is used in excess as a source of fat to increase the ration's calorie density, fibre digestion may suffer. Feeding free or unprotected fat above 1% level has a depressive effect on rumen cellulolytic microbial activity. Unprotected fats produce physical and chemical changes in the microbial fermentation of feed that are generally negative. Because milk fat contains more saturated fatty acids and less polyunsaturated fatty acids, which are atherogenic and increase the risk of cardiovascular disease by raising plasma cholesterol and low density lipoproteins, milk fat is blamed for this (LDL).

Advanatges of supplementaion of bypass fat to dairy animals

- Keeps animals in positive energy balance.
- Increases milk yield
- Increases milk fat content.
- Maximizes peak milk yield and lactation days.
- Improvement in reproductive performances.
- Prevents post-partum weight loss.
- Improves general body condition.
- Protects from heat stress.

Effects are not similar in all types of dairy animals. Effect tends to be greater in high yielding dairy animals. Effect tends to be greater in early lactation (0-90 days) than mid lactation (91-150 days). Effects tend to be more in Holstein cows. Bypass fat to be supplemented at rate of 15-20 g/kg milk production/ animal/ day. It can be supplemented to dairy animals 15 days before to 150 days after parturition. Bypass fat can be supplemented to dairy animals by mixing with concentrate mixture. It can be given as single dose or in divided doses.

13

Housing Management

By minimizing the harsh climate effects like heat, cold and rain, it is crucial to offer the proper circumstances for raising dairy cows in the tropics. Stress can be reduced through the design of the farm and the dwelling. Controlling the environment lowers the risk of stress and disease, improves management, and increases milk production. Housing can be a key factor in determining productivity when combined with an effective herd health management programme.

Following the guidelines listed below should guide housing site, design, and construction:

Location: Fertile land, the lack of evidence of flooding during the wet season, ease of transportation, proximity to the milk collection centre, availability of supplies for farming, clean and sufficient water year-round, and distance from factories or other farms that may release unpleasant odours or result in pathogenic infections are all factors that need to be taken into account when choosing land for livestock farming.

Ventilation: Housing should be located in clear or on high ground to allow for improved ventilation. The roof must have a height that allows heat, moisture, and pollution to escape. The stable's sloped roof can shield it from the impacts of the sun and rain, and breezes can pass through it to keep the home cool and dry.

House facing: Housing is advised, with the front of the home facing east-west with a gable roof. The farm's longitudinal axis ought to be aligned with the sun. The shed should be shielded from morning and afternoon light as well as rain by being positioned in this direction.

Sanitation: The milking space should be built with ease of work, milking, feeding, and cleaning in mind.

Roof: By enabling air to flow through the roof and promoting ventilation, good roofing protects against the hot climate. Animals need to be protected from cold stress, and from heat loss at night, so the roof or shelters must be well-designed. In the tropics, thatched roofing is widely utilized since it is affordable and, when constructed properly, may offer good protection. Where they are less expensive, tiles are favoured.

Walls: Housing in the tropics is typically open-sided to maximize ventilation and air movement through the shelter to maintain a temperature-consistent interior environment.

Floor: The floor must be made of concrete or a material that is simple to clean. It should be constructed with appropriate drainage in mind and adhere to hygienic standards. Faeces and urine should be constantly drained or removed, and the floor should slope and have channels that allow urine to run away. In order to reduce pathogenic microorganisms and the amount of flies, homes should be routinely disinfected.

Dairy shed: Housing for smallholders should be constructed to maintain concentration in the milking area. Before or during milking, farmers should utilize feed concentrate and/or roughage in the milking stable. Equipment for milking should be stored in a shady place.

Foot bath: Footrot can be decreased by creating a place for dipping feet before entering the milk stable and/or after milking. An appropriate disinfectant is clean water diluted with three to five percent copper sulphate.

Farmers should have well-designed housing with sufficient air movement and shade to lessen instances of heat stress. Animal handling should be avoided during times of excessive humidity and warmth, and good-quality, cool water should always be available. Having a place where cattle may graze and exercise benefits health. Exercise or rest places ought to be near the milking shed. Similar concepts apply to calf housing, which has raised slatted floors for good ventilation. To prevent calves from suckling each other, ingesting hair that results in hair balls, which then induce bloat in calves, it is best to keep them in separate housing. This also guarantees that they get their daily food allowance.

14

Record Management

Dairy farmers hope to have better milking cows than those in their current herds. A solid foundation is necessary for every farming operation. Undoubtedly, the dairy farm documentation system is among the most important facets of farm management. Tracking each cow's productivity within a herd and assisting breed development generally are the major objectives of a dairy farm tracking system. This information could be referred to as production data. A cow has two "calving dates": one to mark the start of lactation and the other to mark its end. These are used to calculate lactation production and can be regarded of as "cow event data." By collecting additional "cow event" data, such as identified heat dates, mating dates and sires used, calving difficulty data, and health events like mastitis, feed problems, and reproductive problems and their treatments, a very useful database of data is created to help farmers better manage their herds. It also offers market data that veterinary field staff may use to identify health and management issues, which will aid farmers in boosting the productivity of their cows. Maintaining a herd recording platform and service that is user-friendly and pertinent to the requirements of dairy farmers is essential.

Data collection and Information recording

Detailed pedigree information about each animal's sire and dam as well as the gathering and storage of data on each animal's unique identification is necessary for successful animal breeding. Little genetic change in a population can occur without these two pieces of knowledge. Today, animal monitoring requires the ability to identify animals. Keeping track of and registering animals, but it is vital to do so in order to improve the breed or population. Making sure that as few errors as possible enter these databases requires a lot of work. A lot of data are now gathered and sent to recording centres electronically, which has eliminated many inaccuracies or discovered faults at the farm level that might be remedied immediately. Data collection has also benefited from the use of computer systems on farms. In order to manipulate data and perform data analyses on computers, the records and pedigrees must be electronically saved. Jay L. Lush started demonstrating to people how data could be used to find genetically superior animals, primarily dairy bulls, in the 1930s. Through the years, the statistical methodology has improved, particularly because to Charles R. Henderson's work between 1950 and 1989. Advances in computing power, such as more memory, more disc space, faster CPUs, and parallel processors, make improvements conceivable. It is ideal to record every animal in a herd without making any decisions about selecting animals to record.

Farm records are kept for all or some of the following reasons:

- To assist in financial planning decisions
- To control labour
- To assist in land management decisions
- To assist in livestock management decisions
- To select the best cows and bulls to breed better cows and bulls
- To assist in breeding, culling and other decisions.
- To evaluate genetically
- To plan future actions

Animal breeders use statistical models to examine the data and assess the breeding potential of each animal in a herd. The estimated breeding values (EBV) are used to rate the animals, and the better ones are mated while the worse ones are eliminated. In order to account for the heritability of each characteristic and the genetic correlations between the qualities, animals are typically judged for a number of attributes that are then weighted by their respective economic values. An EBV includes information from (1) records about the animal, (2) documents about the animal's sire and dam, and (3) records about all of the animal's offspring. At the same time effects due to the herd, the year of birth, the age of the animal and many other factors need to be removed during the estimation process.

Record should satisfy the following criteria

- They need to be helpful. Data shouldn't be captured at all unless it will be utilised (converted into information) at a later time to make management decisions.
- Records must be maintained in a format that makes it simple to turn them into information: To ensure that the format in which the data are saved will simplify later analysis and interpretation, the intended end

use must be determined before retaining a record. Too frequently, the end use is overlooked, substantially reducing the data's utility.

- Simple record-keeping procedures are required since dairy farmers already have a lot of works. Complicated record-keeping procedures are also practically impossible to delegate to staff members because they are difficult to comprehend and time-consuming to perform.
- There should be as little duplication as possible. Even while some data may need to be recorded multiple times or in different formats, this must be kept to a minimum. To put it another way, if a record is to be kept in the field, the recording system should be set up so that data may be easily input there rather than having to be entered again at the office.

Records must lead to actions being taken: Unless a record is specifically intended to be used for some future action or in management planning it should not be kept.

Easily updated

Easily understood

Easily summarized

The keeping of dairy records can be divided into the main activities of:

A. Identification of animal

B. Breeding and Calving Record

C. Calf Record

D. Feeding and Production Record

E. Health Record

F. Estrous Detection History

A. Identification of Animal

Dairy farmers can benefit from animal identification in the following situations:

(a) Choosing a sire and wanting to avoid inbreeding;

(b) Evaluating heifers for breeding and needing to select superior bulls;

(c) Assessing overall herd reproduction and needing to know the age of heifers in their first heat;

(d) Knowing the age at which a heifer should be targeted for breeding;

(e) Determining whether an animal is the right size for its age

(f) Selecting the animal to be culled by comparing genetic lines across a region and

(g) Selecting the animal to be culled based on age

A farmer who keeps records of identified cows can better manage his breeding operations by establishing breeding dates and feeding schedules, identifying calves, sires, and dams, determining the time for pregnancy testing, and figuring out when a cow should give birth.

Permanent or temporary identification methods are both possible. Since they can be lost by an animal permitted to exercise or graze in a field, ear tags, neck chains, and ankle tags are typically temporary. Permanent marks are those that cannot be removed and include tattoos, hot brands, and freeze branding. Every form of identification has flaws, such the difficulty of applying tattoos and brands. This makes it possible to employ multiple systems in combination. In cases when animals are kept in stalls, a stall nameplate might be the best record. The cow's name or number, date of breeding, date of calving, milk and fat output, feed type, recent health interventions, and other details are recorded on this, which may be made of sturdy paper or a blackened board.

With the proper tattoo tools, which increase the number tattooed by one with each subsequent application, tattoos may be applied on animals' ears. Holding a light behind the ear that illuminates the darker portions of the tattoo can help with reading since tattoos are difficult to read. Tattoos are especially challenging to read on naturally dark skin. Brands can either be of the hot or freeze variety. Hot branding disfigures a animal's hide yet is permanent and easy to read. This has effects on the look of the animal during its productive life as well as the value placed on the hide.

An extremely popular method of identification is metal ear tags. These are fastened by using specialized application pliers, and despite appearing to be permanent, they could be removed relatively ease if required. Tags could also be challenging to read. Some of the plastic ear tags, which come with their own applicator and are made by different businesses, have shown to be reasonably durable, especially when used in stationary environments. Ankle tags and tail tags are preferred for milking parlour designs where the animal's head is less frequently seen than neck chains, neck straps, or horn chains with numbered tags.

Broadly following are main forms of animal identification

- Physical characteristics include color, horns, size, form, and eye patterns.
- Ear notch: Ear notches help breeders maintain accurate records in regards to registration and in regards to health records on the farm

- Branding: In some nations, especially if hot branding is used, this may be a problem for animal welfare.
- Temporary identification techniques include color markings (with paint or crayon), temporary tags attached using a variety of techniques.
- Visual tags are frequently used on the ear.
- Electronic tags: Radio frequency identification (RFID) chips can be inserted under the skin, on a visible tag, or within the tag itself. There are various types of electronic tags, and before using a certain type, the intended usage must be understood.

B. Breeding and Calving Record

One breeding and calving record is also required. It can be used in conjunction with production data to create a lactation record. Previous calving data for a specific cow may provide insight into the difficulties of calving to be expected during a freshening. Keep a record of the reproductive tract examination findings after calving. Additionally, the health record might contain these data.

C. Calf Record

As soon as the calf is born or acquired, a permanent record should be started. Identification is the key to the entire system, and the optimal time to identify an animal is at birth. Every animal needs to be permanently identified in some way. Temporary identifying devices include ankle straps, ear tags, and neck chains. Information about the sire and dam is required for genetic analysis and for the planning of upcoming breeding. A genetic examination of the sire's ability to facilitate easy calving is influenced by the sex, birth weight, and calving difficulty.

D. Feeding and Production Record

In order to determine a cow's profitability, determine her nutritional needs, and do genetic analysis, it is necessary to keep track of the feed she consumes and the milk she produces. It is not necessary to keep a record for every production day, but during breastfeeding it should be done at least once a month. It should be possible to record weight and lactation number, two factors that are crucial for animal evaluation and ration design, respectively. Although milk can be easily weighed, laboratory testing is required to determine the amount of fat, protein, and somatic cells. The difference between income and feed costs depends on the current feed and milk prices. Somatic cell counts are a very helpful indicator of the severity of mastitis.

E. Health Record

Any disease, damage, and special health examinations the animal has had to endure should be documented in this file. The date should be recorded together with any relevant information about the condition, harm, or test type. The course of treatment, if any, as well as the outcomes and associated costs, including any missed wages from things like milk withholding, should be documented. The criterion for culling can be based on health records. They may also offer hints about upcoming illnesses and information useful for tracking health issues in a herd.

F. Estrous Detection History

Records of estrus for individual lactation should be recorded. This record requires identification, birth date, lactation number and dates in estrus which should be recorded. Estrus records for every lactation should be kept. Identification, birth date, lactation number, and estrus dates must all be entered into this document. Poor estrous detection is known to be the main cause of ineffective breeding. The main goal of the estrous detection chart is to facilitate and enhance estrous detection. The average estrous cycle for cows is 21 days. By adding 21 days to the last day a cow was in estrus, it is possible to estimate the next estrus. If the cycle is longer than 21 days, it is possible to determine a cow's unique cycle length by watching numerous cycles, which helps with the forecast of the following estrus. The likelihood of seeing cows in estrus will increase if the person in charge of estrous detection knows when to anticipate the next estrus and which cows to watch. Estrous detection chart can help with the issue of timely heat detection. A history of estrous detection might help to identify breeding issues. Significant cycle deviations in cows could indicate cystic ovaries or other health issues. It's possible that other cows don't show any overt signs of estrus.

Data processing

Data processing for genetic improvement necessitates a significant amount of computation. The techniques employed depend on the breeding program's nature and the data needed to make informed selection decisions. This section will cover the strategies for ranking animals for selection purposes as well as the procedures for adjusting individual data for environmental impacts.

Environmental factors influencing production

The performance of an animal can be influenced by external and internal factors. External factors affect the whole herd or flock, e.g. herd, year, season, while internal factors affect the individual animal, e.g. weaning age.

Herd, season and year differences: The contemporaneous comparison and herdmate comparison processes are typically used to eliminate herd, year, and seasonal variances. Both techniques are based on individual performance variances from the mean of animals milking in the same herd concurrently. However, the herdmate technique uses age-corrected records of all the animals starting lactation during the year and season under consideration, whereas the contemporary comparison only considers the first lactation records. Contemporary comparisons are ineffective when a herd has few first lactation animals during a given year and season. Therefore, when herd sizes are small, the herdmate approach should be used.

Correction for environmental effects

True genetic ability is frequently obscured by environmental circumstances. In order to reduce environmental variation, performance records should be modified for known environmental variables. The extent of the environmental influences and how precisely they can be evaluated determine the usefulness of this modification. The easiest and most effective method is to compare the individual's performance to the average of other animals exposed to comparable circumstances. This is feasible in huge populations only. The approach has substantial sampling mistakes in small populations. Therefore, correction factors are frequently employed to account for environmental influences.

15

Reproduction Management

Animals need to function efficiently during reproduction in order for a dairy production system to be sustainable. This is necessary to produce milk, the primary product of interest, as well as to supply replacement animals. From the perspective of reproduction, delayed puberty, lengthy calving intervals, limited productive lives, and high calf mortality are the key variables that cause economic losses. Reproductive management's proportionate contribution to the effectiveness of dairy production systems varies greatly. It may differ according on the environment, genotypes, volume of inputs and outputs, the farmers' managerial abilities, and economic as well as societal factors. Additionally, there are significant changes in the reproductive processes and typical fertility parameters between tropical cattle and buffalo depending on the genotype and how it interacts with the environment. Generalizations are therefore challenging due to the numerous genotypes utilized in smallholder dairy production as well as the wide range of conditions and production systems in which they are raised. However, from a practical standpoint, sustaining an ideal level of reproductive efficiency mostly involves matching genotypes to the resources at hand and employing effective management techniques to enable the animals to express their full reproductive potential. Higher reproductive efficiency during the course of a female's lifetime results in more calves for use as replacements or for sale in the herd, as well as more lactations and consequently more milk. As rearing circumstances become more rigorous and more incomes are required to cover costs for labor and feed, these factors become more crucial.

Dairy animals must carry out the following criteria in order to fulfill their duty as an economically practical animal for their owner:

- From birth to puberty grow quickly
- Reach puberty at a young age
- Easily conceive from a fertile mating
- Create a healthy calf.
- Make enough milk to feed the calf and extra to sell.

- Early postpartum estrus re-entry and subsequent conception
- Continue producing calves and milk at regular intervals till the end of its productive life.

The ability of an animal to meet these requirements will depend on many factors.

Puberty

The first behavioural oestrus that is accompanied by ovulation and the formation of a healthy corpus luteum (CL) in the ovary is considered puberty in female ruminants. Numerous factors, both endogenous (such as genotype, growth, and body weight) and exogenous (such as year or season of birth, rainfall, diet, thermal environment, photoperiod, and rearing method) influence this. Puberty typically occurs when cattle and buffalo heifers weigh between 55 and 60 percent of their adult body weight. For cattle and buffalo, the age at which they reach puberty can vary greatly, from 12 to 40 months for cattle and 18 to 46 months for buffalo. Therefore, growth rate and body weight play a bigger role in determining puberty than age. In ideal circumstances, river buffalo and their crosses mature before swamp buffalo, and taurine cattle and their crosses mature before zebu cattle. However, zebu cattle and buffalo typically live longer productive lives than taurine cattle, making up for some of the drawbacks of late puberty. The year or season of birth, genotype, diet, management, thermal environment, parasites, and diseases are the key variables that affect when an individual reaches puberty.

Sexual cycles and mating

The typical oestrous cycle for adult, non-gestational heifers and cows lasts an average of 21 days. Pro-oestrus, oestrus, met-oestrus, and di-oestrus are the four phases of the oestrous cycle. One follicle (sometimes two) from a cluster of developing follicles in the ovary continues to grow and mature while secreting estrogens during pro-oestrus, with the help of the pituitary hormones luteinizing hormone (LH) and follicular stimulating hormone (FSH). The brain of the cow is affected by estrogens, which results in the behavioural modifications associated with oestrus or heat. The vulva may expand, the vagina may become hyperemic, cervical mucus may be secreted, and uterine tone may increase as a result of impacts on the reproductive system. After the heat period has passed, ovulation results from the high oestrogen levels' significant LH release. Following ovulation, the follicle remains are converted into the CL, which secretes progesterone and primes the reproductive system for conception. If conception takes place, the CL continues to release progesterone for the majority of the pregnancy. This stops future cyclic activity and ovulation, however on rare occasions; a small fraction of animals may exhibit some signs of heat. The chemicals prostaglandin F2a (PG) and oxytocin, which are secreted by the uterus and ovary respectively, work in tandem to cause the CL to be destroyed if fertilization is unsuccessful or if the embryo dies before days 14 to 15 of the cycle. As a result, the oestrus cycle can be repeated, giving the animal more chances to get pregnant. The outward indicators of heat often appear less in buffalo and are more evident in taurine cattle than zebu cattle. However, there are noticeable differences across breeds and even within breeds, and the symptoms can vary greatly in strength.

The primary oestrus indicators are

- vulval swelling and reddening
- the vulva secretes a clear, glassy, stringy mucus
- ligaments of the pelvis to relax
- agitation and/or shouting
- decreased milk production and hunger
- a desire to converse with animals
- Remaining still when being ridden by a bull or another cow.
- Males will approach and attempt to mount females who are in heat when they are in the wild. Additionally, females in cattle will interact with and mount one another during heat, although this is uncommon in buffalo. In the latter scenario, frequent small-volume voiding may occasionally be observed.

Tropical cow breeds typically experience heat for 10 hours on average less than temperate types (mean 15 hr). Environmental elements like temperature and humidity, social factors like an animal's place in the herd's dominance hierarchy, and the presence of sickness or uncomfortable conditions in the animal's legs or hooves can all affect how much heat an animal expresses. The ability to perceive heat is crucial for dairy cow and buffalo to reproduce. Visual observation is the most affordable and practical approach. But in order for this strategy to work, the observer must be motivated and knowledgeable of what to watch for.

There are many tools for detecting heat, such as movement detectors (pedometers), vaginal resistance measurements (which decrease at oestrus), vaginal mucus examinations for stretchability and crystallization patterns, body (or milk) temperature monitoring, and repeated progesterone concentration measurements. But in underdeveloped nations with smallholder farming systems, the majority of these are not feasible. In bigger herds using grazing management techniques, the employment of "teaser" animals may be appropriate. These animals include males who have undergone vasectomies. To help them identify the cows they mount, they are typically equipped with chin-ball identifiers.

Ovulation occurs in both cattle and buffalo following the conclusion of the heat cycle, at about 12 hours in the case of cattle and 14 hours in the case of buffalo. Therefore, the last few hours of heat or right after it ends are the best times for mating. This is due to the requirement that male sperm spend at least six hours in the female reproductive system before they are capable of fertilizing an ovum. The sperm can survive in the female reproductive tract for 24 hours, and the ovum can survive after ovulation for 12 hours.

The ovum is fertilized in the oviduct, and the embryo that results reaches the uterus after four days. The embryo grows and divides its cells quickly. Progressive implantation, or attachment to the uterine lining, takes place 25 to 35 days following fertilization. From 45 days following fertilization, the embryo is referred to as a foetus. In Zebu cattle and taurine dairy breeds, the average gestational period is 285 days and 280 days, respectively (range 270-290). The average gestation period for buffalo is 316 days for swamp types and 310 days for river forms (range 300-330).

A veterinarian will typically palpate the genital tract through the rectum 50 days or so after mating as the normal procedure for determining whether an animal is pregnant. The criteria for determining pregnancy in buffalo by rectal palpation are identical to those in cattle, however because to the extended gestation period, each palpable feature becomes apparent for the first time in pregnancy roughly two to four weeks later. Pregnancy can also be diagnosed using other cutting-edge techniques, like ultrasound scanning and hormone levels in milk or blood being measured. The process of calving (parturition) begins at the conclusion of gestation. There are three parts to it: dilating the delivery canal (2–6 hours); removing the foetus (30–40 minutes); and removing the foetal membranes (2-6 hr). Calving should take between 8 and 12 hours in cattle and 6 to 8 hours in buffalo under typical circumstances.

The cow's reproductive system undergoes a phase of healing after calving termed involution, during which the uterus regains its pre-pregnancy size and condition. Normally, this is finished in 25 to 35 days. However, if the uterus contracts an infection after giving birth, this procedure may be postponed. This may occur if the cow gives birth in unclean, filthy settings, or if the baby is delivered abnormally due to dystocia, a retained placenta, or uterine prolapse.

The hypothalamus in the brain, the pituitary gland beneath the brain, and the ovary in the abdomen are the organs that regulate the hormonal mechanisms that control the sexual cycle. These organs gradually regain their cyclic functional status, and the cow should typically exhibit signs of heat 30 to 60 days after calving. The activities of these organs are affected by a variety of variables and as discussed in the following section.

Genotype

Although an animal's genotype dictates its capacity for reproduction, all of the previously listed elements—collectively referred to as environment—have an impact on the final outcome. Therefore, a well-adapted tropical zebu cow may have a relatively low innate reproductive capability, but under the current conditions, it may be able to reach its full potential. If transported to the tropics, a Holstein Friesian (HF) cow that was bred for high milk production and outstanding reproductive performance in temperate climates will perform far below its potential. If their genotype has a large component of taurine blood, even locally bred cows may not be able to adapt to the climate in high temperatures and humidity. A study of dairy cows on smallholdings revealed that as the proportion of HF inheritance increased, the interval from calving to first AI increased and the first service conception rate decreased. This continuous upgrading of the native cattle to HF has been taking place there through an effective AI service. The longer time since the initial AI suggests that the climatic and/or nutritional challenges in these animals with higher HF inheritance make it more difficult for the ovarian activity to resume. Given that unadapted genotypes exposed to high ambient temperatures are known to experience greater rates of fertilization failure and/or early embryonic death, the reduced conception rates in these could be explained by these.

Nutrition and environment

The age at which a person reaches puberty is significantly influenced by nutrition. Heifers that are well fed will grow more quickly and reach puberty earlier than those who are poorly fed for a given genotype of cow or buffalo. As a result, the first calving occurs earlier, bringing financial benefits to the farmer in the form of more milk and more calves during the animal's lifetime. Early lactation in dairy cattle is particularly stressful because the cow cannot get all the nutrients she needs for maximum milk production from her daily feed intake. In order to maintain a metabolic equilibrium, it uses up its fat reserves of nutrients until the second or third month of nursing. Therefore, compared to cows calving in good condition, those in poor condition lose more weight and take longer to recover it back. Therefore, these cows also require longer than well-fed cows to start cyclic ovarian activity.

It is generally known that maintaining normal fertility requires a diet that is balanced and contains enough protein and energy. Infertility has been linked to shortages in sodium, phosphorus, and selenium, and reproductive issues can also result from iodine, manganese, and zinc deficiency. One of the essential vitamins is beta-carotene. Animals kept only on cereal straws or stovers may become deficient in b-carotene, which can cause quiet or delayed ovulation, a poor pregnancy rate, and significant embryo mortality. High temperatures and humidity in tropical environments can have significant reproductive inhibitory effects. This can have a direct effect by reducing appetite and feed intake, or it can have an indirect effect by changing the type and amount of feed that is accessible to the animals.

The ovarian activity in the cows is stimulated by the favourable nutritional conditions, as a result of an abundance of high-quality forages following the monsoon rains, which in turn leads to the seasonal conceptions. In contrast to animals who give birth at other seasons, which endure a lack of feed and heated weather conditions, animals that give birth during this season benefit from improved feed during the time after calving and hence resume sexual cycles sooner. This suggests that breeding patterns may be modified to account for seasonal fluctuations in feed supply and climate, or those animals due to calve during challenging times should get supplemental feeds, thus increasing reproductive efficiency. In the early postpartum phase and two months before calving, supplements should be given.

Managemental practices

The continuation of sexual cycles after calving might be influenced by management techniques like nursing calves. Six distinct suckling regimens were examined in later research: free suckling, free suckling followed by weaning at days 45, 60, or 90, once a day starting on day 7, and twice a day starting on day 7. According to the findings, free suckling had the lowest pregnancy rate at 150 days. The maximum pregnancy rate was achieved by weaning at 45 days, although calf mortality was high. To find the best method for enhancing reproduction without compromising calf growth or survivorship, similar experiments under other smallholder systems where suckling is practiced should be conducted. In addition to the suckling stimulation, studies have demonstrated that the mere proximity of the calf to its dam as well as the degree of direct contact with her through touch, scent, and sound can postpone the start of postpartum ovarian activity in the dam. Substances like b-endorphin and insulin-like growth factor are thought to mediate this. Therefore, altering the calf feeding system to limit the interaction between the cow and its calf is another way to increase reproductive efficiency.

Breeding management

The management choices made by farmers, along with their knowledge and expertise, has a significant impact on the reproductive efficiency of cows. The age and/or weight at which heifers are mated, as well as the time period following calving, must be determined by the farmer. Mating too late will result in lost production time in the case of heifers, whereas mating too early will result in a lengthier postpartum anoestrus period after the first calving. Full fertility may not be attained until about 60 days after calving in cows with normal calvings and uncomplicated uterine involution, even if the cows show heat at any time after 45 days. In many systems, mating delays of more than 60 to 90 days can result in financial losses. After calving, some smallholder farmers frequently wait three to six months before mating. This is because it's thought that mating too soon can shorten lactation and reduce milk output. These farmers need to be educated on the benefits of early mating, such as increased lifetime production of milk and calves by a single cow.

Poor heat detection and improper AI are two significant flaws in many smallholder systems. Farmers will only be able to detect 30 to 50% of ovulations that take place during the postpartum period. However, between 15% and 20% of the farmers report may be incorrect. Therefore, oestrus detection efficiency and accuracy both need to be improved by instruction and training. Although the timing of mating is crucial for conception, farmers may have little control over this. For instance, access to stud bulls can be restricted or AI services might not be widely available. Farmers' propensity to presume that a cow will be pregnant after being serviced is another factor in low fertility. They consequently fail to look for returns to service and discover the cow is not pregnant many months later, by which point they have already lost precious productive time.

Reproductive indicators are a collection of parameters that can be used to evaluate the effectiveness of reproduction. Each index offers data on a certain fertility-related topic, but each has unique value and restrictions. Therefore, a combination of these indices must be utilized in order to estimate the total reproductive performance at a specific time or over a longer period of time. While certain indices can be used on individual animals, others are only appropriate to a herd or community of cows. The age at which heifers reach puberty and the age at which they have their first calf are crucial indicators in this scenario. The former is influenced by the moment when ovarian activity starts, whilst the latter is affected by the moment of conception. Heifers will typically get pregnant shortly after adolescence if they are allowed to roam freely and have access to bulls. The effectiveness of heat sensing, timing of service, and other pertinent aspects, however, will have a significant impact on the age of first calving in constrained systems.

The conception rate (CR) is the proportion of animals that conceive in comparison to the total number of animals used. When compared to all breedable females in the herd, the pregnancy rate (PR) is computed as the proportion of animals that become pregnant. A given year's calving rate is the proportion of breedable females that do so. The number of services per conception (S/C) is calculated by dividing the entire number of services provided to a group of ovulating cows by the total number of conceived cows. In the livestock farming system, the bull, artificial insemination (AI), and cow-related factors all affect these indicators.

The calving interval (CI), whether used to a single cow or to the entire herd, may be the one statistic that offers the most data on reproductive efficiency.

There are three parts to this:

1. Time between calving and the first oestrus (postpartum anoestrus period)
2. Time between the first oestrus and conception
3. Time between conception and calving (gestation period)

Under contemporary intensive dairy systems, it is generally agreed that the CI should be close to one year in order to preserve the best economic benefits. A cow must become pregnant by 80 to 85 days following calving as the normal gestation period is 280 to 285 days. The cow must exhibit heat early in the postpartum period, begin ovarian activity early, conceive easily, bring the pregnancy to term, and give birth to a calf. The farmer must recognize heat, mate the cow at the appropriate moment, and supply enough food and other inputs. In the various smallholder farming systems, these characteristics' relative importance varies. For instance, nutritional changes brought on by seasonal herbage shortages under broad free grazing circumstances delay puberty and the postpartum cycle. However, once ovarian activity starts, cows that are grazing with intact bulls quickly become pregnant. Factors like heat detection, mating time, or AI quality are especially crucial in constrained or tethered systems. In some smallholder systems, the presence of the calf and the frequency of nursing also affect postpartum ovarian function.

A one-year CI is frequently difficult or impossible to obtain under many tropical dairy systems, and in other cases, it is even undesirable. A percentage of cows calving only once every two years are typical in many traditional dairy operations. Similar to this, many tropical smallholdings should not employ the acceptability standards used for the other reproductive indices under temperate conditions. For such systems, a more feasible and realistic set of criteria needs to be created. It then needs to be validated under the current production and pricing levels. The table 10 lists the traditional dairy cattle reproductive indices at their optimal levels, along with a suggested list of "acceptable" standards for usage in more advanced smallholder circumstances. In order to account for intrinsic differences, such as the later maturity of heifers and the longer gestation period, a comparable set of criteria is also provided for buffalo.

Table 10: Reproductive indices for dairy cattle and buffalo under optimum conditions and suggested `acceptable' performance under improved smallholder systems

	Cattle		Buffalo	
	Optimum	Acceptable	Optimum	Acceptable
Age at puberty (m)	< 18	< 24	< 30	< 36
Age at first calving (m)	< 30	<36	< 42	< 48
Calving to first service (d)	< 60	< 90	< 60	< 90
Calving to conception (d)	< 85	< 115	< 85	< 115
Calving interval (m)	12-13	13-14	13-14	14-15
First service conception rate (%)	> 60	> 55	> 55	> 50
Overall conception rate (%)	> 80	> 75	> 75	> 70
Calving rate (%)	> 75	> 70	> 70	> 65
Services per conception	< 1.6	< 1.8	< 1.8	< 2.0

Between various conventional systems, there are significant differences in reproductive effectiveness. For instance, in Zebu cattle, the CI ranged from 12.2 to 26.6 months, the first-calving age was between 23 and 58 months, the calving rate was between 20 and 90%, and the number of services per conception was between 1.4 and 2.8. It is obvious that the values given in Table won't be possible in the harsher situations. However, many smallholder systems can easily reach these goals while still being cost-effective because to the adoption of the right genotypes, deliberate dietary adjustments, and improved reproductive control.

Infertility and sterility

Sterility is the total and permanent inability to produce offspring, whereas infertility refers to a diminished ability or temporary inability to reproduce. The underlying reasons of infertility or sterility in an individual animal or a herd must be correctly diagnosed. Only then can a logical choice be taken in order to solve the issue and avoid a repeat of it.

Finding the cause of infertility or sterility necessitates:

- background information on the environment
- examination of the farm's operational characteristics
- complete history of the current issue (on an individual and herd basis)
- complete clinical evaluation of the concerned animal or animals or the herd
- special diagnostic procedures as needed.

A choice must be made regarding the best corrective actions when the issue has been recognized. These could include adjustments to management, nutrition, or breeding practices, as well as the care of the affected animals and follow-up steps. The method chosen should take into account factors including practicality, economics, sociology, and culture.

Delayed puberty

The replacement heifers are frequently given relatively little particular care under smallholder management settings, which causes late puberty, a delayed first calving, and financial loss for the farmer. This is especially clear when artificial insemination (AI) produced calves with a high amount of taurine heredity are housed in the same stressful conditions as native calves, with little nutrition and no parasite or disease management. When implementing such genetic improvement initiatives, the AI and extension services must make sure that smallholders are aware of the extra inputs needed to reap the rewards of rearing superior genotypes and are actually capable of providing them. Congenital defects that delay or entirely prevent reaching puberty may occur in a small number of calves. Such animals ought to be found as soon as possible and culled. Additionally, because many of these ailments are inherited, such animals shouldn't be used as breeding stock. Ovaries that are nonexistent, aplastic, or underdeveloped are examples of congenital ovarian disorders. The heifer will be sterile if the disease is bilateral, however in unilateral situations, ovulation can take place from the normal ovary. However, since the condition will be passed down to succeeding generations, such animals should never be employed as breeding material. Rectal palpation of the ovaries might be used to diagnose the disease. When twins of different sexes are born in cattle and buffalo, the female calf is typically infertile. The reproductive organs of the female twin have been altered as a result of mixing blood from the male twin during early pregnancy, a condition known as free-martinism. Ovo-testes, an underdeveloped female genitalia, and occasionally the formation of male accessory glands are the results of the male twin's inflow of primordial germ cells and testicular hormones. A clitoris with good development and a hair tuft at the ventral vulval commissure are possible external features. The vagina is typically smaller than it should be, and rectal examination can confirm the diagnosis.

Improved nutrition and care are advised if a complete clinical evaluation reveals no other problem besides dormant ovaries. The cow has likely started the sexual cycle but the farmer missed seeing heat if rectal examination confirms the presence of a CL in one of the ovaries and one or more follicles of different sizes. Advice on how to improve heat sensing should be given in such circumstances. It should be mentioned that hormone treatments to accelerate puberty in heifers with delayed puberty are not advised. This may lead to the transmission of unfavourable features to the following generation and a decrease in herd fertility.

Anoestrus

Anoestrus is defined as the lack of oestrus symptoms. It can happen under circumstances that are consistent with normal physiology, such pregnancy and the first few weeks after delivery. It is regarded as a cause of infertility when it happens when breeding is necessary, which is typically at the anticipated time of breeding after reaching puberty in heifers and, depending on the genotype and production method, after 45 to 90 days following calving. The underlying causes of anoestrus include an abnormally long period of postpartum ovarian inactivity, also known as "true" anoestrus, cystic ovaries, uterine infections, pregnancy problems that cause the CL to persist, or non-expression by the animal (silent ovulation), and the farmer's failure to recognize heat. If the situation is real anoestrus, ensuring that diet and management are satisfactory should come first. Modifications that are compatible with the management system should be put in place if the calves are being suckled or tied near to the dams. Hormone therapy has erratic outcomes and should only be tried in cases where the body is in good shape, the reproductive system is healthy, and the ovaries are normal in size and show signs of having little tertiary follicles. There are two options available. The first involves either simulating FSH using Pregnant Mare Serum Gonadotrophin or inducing FSH and LH secretion by administering Gonadotrophin-Releasing Hormone (Gn-RH). The second strategy involves using the so-called "re-bound" effect, which involves giving exogenous progesterone for 10 to 12 days before abruptly stopping. This can be done with silastic devices impregnated with progesterone that can be put in the vagina or implanted under the skin of the ear, or orally active chemicals, which are inexpensive but impracticable in many instances. Treatment is necessary for ailments including luteal cysts and uterine infections. However, the latter illness may also need to be treated with the proper antibiotics. Early cases of both conditions typically react well to the luteolytic effect of prostaglandins (PGs). Long-standing cases, however, can relapse and need to be culled for financial reasons. PGs can also be used to effectively treat mummification, although maceration typically leaves the uterus permanently damaged and reduces fertility in the future. It can be challenging to tell whether a cow has silent ovulation or is sub-estrous when evidence of ovarian activity is discovered in animals that the owner claims are an-estrous (missed estrous). It is frequently possible to estimate about when the next heat might be anticipated based on the results of rectal palpation, and the owner can then be given instructions on what to watch for and when. If this too is unsuccessful, it is feasible to stimulate estrous by injecting PGs while a mature CL is present, focusing heat detection over the following three days, and then mating at the proper moment.

Repeat breeding

Repeat breeding is the inability to conceive after receiving three consecutive services during 'normal' oestrous cycles without any obvious anomaly. Repeat breeding can have a variety of root causes, including those involving the cow, the bull, the AI service, and farm management procedures. After one cycle length away from service, a cow returns to heat. Congenital abnormalities or acquired abnormalities (occlusion of the oviducts due to injuries sustained during calving or scar tissue formation after infections); and ovulatory disorders are the main factors affecting females (delayed ovulation and non-ovulatory heat). The bull's fertility, the time of service in respect to heat, and the number of services during one heat phase are all relevant elements in the case of natural service. When artificial insemination is employed, factors such as the semen's original fertility, processing, storage, and handling, scheduling of the service, and inseminator expertise all play a significant role.

Early embryonic death (EED), which is defined as the death of the embryo before days 14 to 15 of the cycle, is the second cause for returning to heat after one cycle length. As a result, the reproductive system behaves as if the animal did not conceive and releases the luteolytic hormones that kill the CL, resulting in heat at the typically anticipated period. Such death occurs prior to the mechanism of "maternal recognition of pregnancy" being engaged. On the other hand, pregnancy recognition has already taken place and luteolytic hormones have been repressed in cases of late embryonic death (LED), which happens after days 16 to 17. As a result, the cow returns to heat later since the reproductive system needs more time after LED to re-start the luteolytic action. Atypical fertilization, such as polyspermy brought on by the ovum ageing, genetic abnormalities in the ovum or sperm (lethals), and an adverse uterine environment are the causes of EED. A thorough examination of all potential reasons, both on an individual and herd or population level, as well as the meticulous eradication of any implausible cause, is necessary for an accurate diagnosis of the causes of repeat breeding. The decision of whether to treat the cow must be made if a clinical anomaly is discovered in the animal. If management or breeding practices are found to be deficient, these should be rectified. However, it is frequently impossible to provide a conclusive diagnosis. The ideal course of action in these situations is to keep the cow with a reliable bull during a period of normal heat or, if that is not possible, to repeat artificial insemination (AI) using high-quality semen and a knowledgeable inseminator. In the absence of a certain diagnosis, indiscriminate hormone or pharmacological treatment usually just serves to exacerbate the issue.

Infectious causes

Specific illnesses or infections without a known aetiology can lead to infertility. It is possible for certain infections to be bacterial, viral, or protozoal. Although brucellosis is a bacterial infection that can affect animals of any age, it only causes abortions in the final three months of pregnancy when it emerges. By coming into contact with or ingesting aborted material, as well as occasionally through milk, infection spreads. It can start a "storm" of abortions when first introduced to a herd. Abortions will afterwards be uncommon, although related conditions including infertility, stillbirths, retained placentas, and others will continue to exist. Man-to-man transmission of the illness results in an illness known as undulant fever. Simple plate or tube agglutination tests, more complex techniques like the complement fixation test or the enzyme-linked immunosorbent assay, or blood tests (for individual cows) are used to diagnose the condition (ELISA).

Another bacterial condition that can affect men, leptospirosis, also results in late-term abortions in pregnant women. Spread happens when an organism that is excreted in an infected animal's urine comes into touch with, or is consumed. Blood and urine tests can be used to get a diagnosis, and antibiotic therapy can be used to treat the illness. A bacterial infection known as campylobacteriosis (vibriosis) is frequently spread during mating. Trichomoniasis, a disorder brought on by a protozoal organism, is another highly comparable ailment that is also acquired venereally. In the first three to four months of pregnancy, both disorders cause the embryo to die. Uterine infections frequently follow this and can last for a variety of times. Bulls can catch the illness from cows when they mate. Serological testing or organism culture can be used for diagnosis. The recovery process is aided by proper treatment of sick cows and two to three months of sexual abstinence, but bulls should generally be culled because they are often asymptomatic carriers of the disease. Bovine herpes viruses 1 and 4, bovine virus diarrhoea, and parainfluenza 3 virus are the primary viral infections that cause distinct reproductive issues in cattle. Through unsuccessful fertilization, the death of embryos, and abortion, these can result in infertility.

Most commonly, germs that enter the reproductive tract through the cervix cause non-specific infections. This frequently happens when unsanitary conditions predominate during calving, natural mating, or artificial insemination. Endometritis will first manifest, and it can develop into metritis and pyometra later on. Animals with endometritis typically cycle, and the only visible symptoms may be white streaks of pus in the clear mucus released during oestrus; nonetheless, these animals will have low fertility due to EED or non-fertilization. Due to the existence of the CL, an-estrous is typically present with metritis and pyometra. Depending on whether the cervix is closed or open, pus discharge might be nonexistent, negligible, or profuse. Inducing estrous is the most effective way to treat uterine infections because the physiological effects of oestrogen on the uterus result in the evacuation of its contents and the activation of its built-in defences. Injecting PGs can be used to accomplish this when a CL is present.

Accidents and miscellaneous causes

After fertilization, there are many different reasons why an embryo could die. Important causes include genetic anomalies in the embryo, environmental pressures, other stresses, and specific or general infections. Few of pregnancies result in foetal death, even under ideal circumstances. The main reasons of death for an embryo after 45 days, when it develops into a foetus, include genetic anomalies, extreme stress, accidents, and particular infections. Following foetal death, the foetus will either be retained in the uterus or will be expelled (abortion). Abortions shouldn't account for more than three to five percent of pregnancies under normal circumstances. To ensure that infectious illness is not present, all cases of abortion should be recorded so that investigations can be conducted.

Male fertility

Males go through puberty gradually, increasing their sperm production and ability to reproduce. Many temperate breed bull calves exhibit libido before turning a year old, although fertility may not be reached for another 14 to 16 months. Bovines need at least 50 million sperm each ejaculation, with more than 10% of them displaying increasing forward motility, in order to reach puberty, which is typically described as the age at which a male becomes capable of impregnating a cow. Similar to in females, nutrition, genes, and many other factors can affect when a calf reaches puberty. In juvenile bulls, the amount of semen generated per ejaculate ranges from two to five millilitres to five to fifteen millilitres in mature bulls. In a typical sample, one to three billion sperm should be present per millilitre, with more than 60% of the sperm being alive and exhibiting robust motility. A sample of semen can be evaluated using a wide range of specialized assays, including microscopic, biochemical, and computer-based procedures. If maximum fertility is to be achieved, the procedure of choosing, raising, and using breeding bulls is crucial. It also depends on whether they are intended for AI or natural breeding. Breeding bulls, however, must always be superior in terms of both their genetic potential and their reproductive traits. They need to be able to breed females effectively, generate excellent-quality semen, and have normal and well-developed reproductive organs.

Natural breeding

Bulls can be employed for one of two different natural breeding methods: regulated hand-mating or free mating on the range. In the old system, the bull detects heat, and cows in heat are often mated multiple times throughout each heat phase. If there is no obvious periodicity in the occurrence of heat, one bull can cover 40 to 50 cows annually. Since it is frequently impractical to introduce two or more bulls at the same time due to hostile behaviour towards one another, numerous bulls may need to be employed in rotation in big herds. In hand-mating systems, the farmer is responsible for heat detection and time of service, and each cow is mated once or twice during each heat phase. A bull can be utilized on three to four cows per week or 150 to 200 cows per year in this case. The first ejaculate is frequently of low quality when a bull is utilized after a break from sexual activity longer than two weeks, hence a repeat mating should always be performed after a few minutes.

Artificial insemination (AI)

Artificial insemination was one of the earliest reproductive technologies used in animal breeding, and it is still the most crucial in many livestock production systems in temperate and tropical countries. One bull can produce enough semen for more than 50,000 cows annually with AI since one bull's ejaculate can supply 400 to 500 cows. The best bulls in the world can be chosen and utilized on cows that are widely separated in time and place preservation methods. Farmers can also obtain a huge variety of bulls and avoid the costs and risks associated with breeding bull rearing. Artificial insemination (AI) can also be used to control several infectious reproductive illnesses. AI, however, has a number of drawbacks. Establishing and maintaining AI centres, together with the associated hardware, people, and training expenditures, are expensive. A strong infrastructure is needed, including a system of AI points, a semen distribution system, field inseminators, and, if frozen semen is utilized, a steady supply of liquid nitrogen. Additionally, a dependable method of communication with the AI service should be in place, and the farming community has to be educated on heat detection and service timing. Transmission of undiscovered genetic disorders or diseases is a potential risk.

The preservation of semen can be accomplished in a number of ways. Deep frozen semen is the most popular; however other varieties of semen can be utilized successfully in many tropical nations where the infrastructure is not sufficiently developed. For preservation, the semen is diluted in a synthetic "semen extender" that includes a number of additives, including chemical buffers (phosphate, citrate, tris), agents that prevent freezing damage and cold shock (egg yolk, milk, coconut milk), sources of energy (fructose), and antibiotics. Each dosage of insemination will include 7 to 30 million motile spermatozoa, depending on the preservation method. Several variables can affect the fertility results of AI. The bull's level of fertility, the unique qualities of the individual ejaculate, and the techniques utilized in its processing, packing, and storage are few important factors. The methods employed for transport, storage, and handling in the field will also have an impact on the quality of the semen after it has left the AI centre. A specific dose of semen may be rendered wholly ineffective by a weakness in any one of the links in this sequence of events.

The inseminator will ultimately determine whether an AI is successful. Before becoming certified, all AI technicians must complete a specialized training programme. However, a variety of individual factors, including innate dexterity, and the frequency of AI, also affect the fertility rate attained. Technically speaking, the handling of the cow's vaginal tract, the processing of the frozen semen, and the proper positioning of the frozen semen are the areas that require extra care.

Modern reproductive technologies and potential applications

Breeders of livestock now have access to a wide range of contemporary reproductive technologies, which have transformed the industry in industrialized nations. While some of these technologies can be employed successfully in tropical smallholder dairy systems with little adjustment, others must first undergo a thorough review and adaptation to the local environment.

Oestrus synchronization

Making several cows go into heat at the same time is known as synchronizing their oestrus. This enables expanded use of AI and better planning of breeding activities. Cows utilised in these programmes must not be lactating and must have regular sexual cycles. The plan, which focuses on regulating the luteal phase of the oestrus cycle, can be carried out in one of two ways: either by prolonging the phase with exogenous progesterone or by ending it with PG or its analogues.

In the first approach, silastic devices carrying the hormone progesterone are inserted into the cow's vagina or under the skin of the ear for 12 to 14 days, along with a dosage of oestradiol at the start of the treatment. The latter approach uses two PG injections spaced 10–12 days apart. Animals often go into heat two to three days after the treatment regimen. Each cow requires two inseminations, either 48 or 72 hours after the progesterone medication is stopped. The PG approach is employed in some management systems in a variant where the initial dose is given to all cows, heat is evaluated, and those showing indications are served again, separated by a 24-hour period. Only those who had not previously displayed heat were subsequently given the second dose, and they received two doses without heat detection at 72 and 96 hours. Smallholder systems have adopted this technique, with varying degrees of success. Success depends on the health, fitness, and cycle of the animals chosen for treatment, as well as their lack of stress and rough treatment throughout AI. Additionally, the treatment schedules and timings should be strictly followed without taking any short cuts out of expediency.

Hormone assay

Techniques like radioimmunoassay (RIA) and enzyme immunoassay (EIA) can be used to determine the concentration of progesterone, which is present in very small levels in females' milk and blood. In the ovaries of cattle and buffalo, an active CL is indicated by a relatively high concentration at a given period, whereas baseline concentrations denote the absence of an active CL. As a result, it is a helpful tool for keeping track of ovarian activity, but it requires repeated samples to find out when, for instance, adolescence starts or postpartum ovarian activity starts. By examining a sample taken between 20 and 23 days after AI, it is possible to identify pregnant animals. A basal progesterone level at 20 to 23 days after AI is an almost unfailing predictor of non-pregnancy because the CL will regress about 16 to 17 days following the last heat if the animal does not become pregnant. The progesterone test has the advantage of allowing non-pregnant animals to be correctly detected at an early stage, allowing for measures to be taken to closely monitor them for heat so they can be mated again at the proper time. High progesterone animals should be kept an eye out for re-entering heat and, if not, should have a rectal exam to confirm pregnancy 60 days or so after mating. The primary drawback of the test is that RIA and EIA tests must be performed in a laboratory using specialist tools. Therefore, widespread application of these techniques is prevented by the logistics of getting the samples to the laboratory and the results back to the farm. Simple cow-side tests on milk have been developed utilizing dip-sticks and strips based on a color shift that denotes low and high progesterone levels in an effort to solve these issues. They have not yet been tested for stability and dependability in tropical environments, and they are still too expensive for routine usage in the majority of developed nations. Future research on low-cost, accurate qualitative cow-side progesterone assays will likely be beneficial for increasing reproductive effectiveness in many smallholder dairy production systems across the world.

Sexed semen

A device called a flow cytometer can differentiate sperm cells that result in a heifer calf or bull calf with over 90% accuracy. The process is relatively slow and also damages sperm cells, but not as much as other methods that have been used to separate sperm. Currently fewer sperm cells are available for insemination and lower pregnancy rates are the two main barriers to this technology's widespread adoption. These issues will be resolved, and sexed semen will eventually be affordable. The first usage will be to raise the birth rate of heifer calves, which might be sold as replacements or utilized to increase the herd. Along with the other advantages, raising heifers at farm would be the most bio-secure approach to increase the herd. The surplus heifers could be sold and provide additional money for those who want to keep the herd size at a certain level. Since heifers are more fertile than cows, investing in heifers with semen always results in a higher return. Getting more heifer calves out of heifers will accelerate genetic advancement. Additionally, when heifers give birth to heifer calves, calving is less challenging. Therefore, increasing the proportion of heifer calves among heifers will lead to more live calves and more first-calf heifers having successful first lactations.

Multiple ovulation embryo transfer

Multiple ovulation embryo transfer, or MOET, was first used by Nicholas and Smith (1983) to refer to the use of embryo transfer and related technology to maximize the genetic improvement of cattle. This entails using elite female donors to provide embryos that are then implanted in recipients (surrogate mothers). Compared to what would be feasible naturally, it enables these superior animals to have many more calves during the course of their lifetime. Multiple ovulation embryo transfer (MOET), oocyte recovery after slaughter, and in vivo ovum pick-up technology are some of the current embryo transfer modifications. Through a hormone therapy regimen referred to as super-ovulation, the donors in MOET are encouraged to shed many ova instead of just one. After being inseminated with superior bull semen to fertilize the eggs, the donor's uterus is used to collect the embryos after five to six days. Each healthy embryo is either transferred immediately to a recipient or kept in liquid nitrogen for later use after being checked under a microscope. Any abnormal or underdeveloped embryos are then removed. Although some industrialized nations employ ET for commercial purposes, the greatest benefit for tropical nations may be the ability to import frozen embryos rather than live animals. When foundation breeding stocks must be established on public or other institutional farms, this may be appropriate. It is unlikely that ET will soon have any direct uses in smallholder systems.

The majority of MOET plans call for one or more sizable nucleus herds. Young bulls employed in natural breeding would be a more practical way to spread the genetic improvement to the wider populace than embryo transfer or artificial insemination. The MOET process makes use of cutting-edge technology, which initially seems inappropriate for less developed nations. Animals become genetically improved as a result of the following actions.

- Accurately classify genetically valuable animals so that the best can serve as parents for the following generation. Performance testing, progeny testing, and pedigree analysis can all be used to accomplish this. Testing for performance measures the animal as a whole, such as growth rate and efficiency, milk output, or the degree of calving difficulties. A partial estimate of genetic value is obtained because such performance has some genetic component. Performance testing has several benefits, including low cost, quick data availability, and the capacity to test most or all of the population's animals.

- Use intense selection pressure to ensure that only the genetically superior animals are chosen as parents. Selective breeding methods are used to produce genetically superior cattle, including artificial insemination, embryo transfer, preserving the progeny of only the best cows, and employing only a small number of the best bulls. Cows have low reproductive rates, hence the majority of genetic advancement is achieved by choosing bulls and producing numerous offspring per bull.

- Reduce the generation interval. Genetic progress will be roughly twice as fast if selection steps can be made every three years as opposed to every six. Because data are not available until the following generation, progeny testing increases the generation gap, which frequently negates the benefits of the higher accuracy.

MOET's main objective is to create selections based on performance tests and pedigree analysis in order to reduce the generation interval, in contrast to the current progeny testing methods. Superovulation and embryo transfer increase the level of selection on the female side. Genetic progress is slightly expedited and evaluations of genetic worth are more accurate when embryos are separated according to MOET schemes. This results from utilizing the same genotype for two distinct phenotypic evaluations. All animals are kept in one or more herds under close monitoring, allowing for reliable and unbiased comparisons between them, which improves the dependability of measurements.

The improvement of milk production should be a particular benefit of MOET treatments. For MOET techniques to function optimally without increasing inbreeding more than 0.5 percent every generation, populations of the order of 1000 animals are needed. The best females are gathered, superovulated, and mated to the best bulls to start such a herd. To increase production, the embryos are then collected and divided (or, in the future, cloned by nuclear transplantation so that many identical females result per embryo). Sexing sperm or embryos might enhance this system much more. The best sets of offspring are then chosen to be the parents of the following generation after being compared for qualities like milk output and milk composition.

In order to have embryos available selection has taken place in the fourth or fifth month of the first lactation, embryos may also be removed from heifers and then frozen. With each generation, these processes are continuously repeated. After a few generations, even the majority of progeny-tested bulls, the average genetic value of chosen animals in these nucleus herds will surpass the average genetic value of chosen animals from outside the herd. Therefore, even though they are not progeny tested, bulls that are siblings of the top females in the nucleus herd are chosen as sires for the subsequent MOET generation as well as for the general population because, on average, they have better genetics than bulls that are readily available elsewhere.

Using cloning in another situation is a variation on MOET that might be an option in the future. If a particular clone turned out to be useful, some of its embryos would be preserved in order to produce numerous cloned embryos through serial nuclear transplantation. The population's genetic worth might then be significantly increased by dispersing these embryos among it. The greatest female clones would eventually mate with the best bulls to produce even better clones, allowing genetic advancement to continue. In the absence of contemporaneous genetic enhancement programmes, which demand sophisticated data collection methods, MOET schemes have the potential to be very beneficial.

Embryo splitting

From the two-cell stage until the developed blastocyst stage, embryos can be divided. However, dividing embryos in the pre-morula stage will not be taken into consideration here due to practical limitations, such as the requirement for surgical recovery. The splitting of embryos is done for two basic causes. The first step is to get identical twins, which are important for both research and specific commercial objectives. Increasing production is the second goal. Under exceptional commercial circumstances, two demi-embryos produce around 50% more calves than a complete embryo.

Splitting does not necessarily require sophisticated technology, making it ideal for less developed nations as long as there is a valid use. Numerous techniques for bisectioning embryos have been documented, ranging in complexity from extremely sophisticated to extremely simple. The majority of these treatments involve immobilizing the embryo and cutting it in half. Applying suction to the zona pellucida creating a depression enabling a device that traps the embryo, or making the embryo attach to a surface, are all methods of immobilization. Typically, a thin glass needle or a piece of a razor blade is used for the bisection. The embryo must be divided after the blastocyst stage in order for both halves to get cells from the inner cell mass. There is no need to revert demi-embryos to a zona pellucida beyond the late morula stage. The embryo must be immobilized without suffering harm, the bisection method must not cause excessive cell damage, and the embryos must be bisected sufficiently symmetrically for pregnancy rates to be high.

Somatic cell nuclear transfer

Somatic cell nuclear transfer (SCNT), a laboratory procedure for producing a clonial embryo utilizing an ovum with a donor nucleus, is used in genetics and developmental biology. It is frequently referred to as "therapeutic cloning" and can be employed in regenerative medicine or embryonic stem cell research. It can also be employed as the initial stage in the reproduction cloning process. A somatic cell (a body cell other than a sperm or egg cell) is detached from its nucleus, which houses the organism's DNA, and the remaining portion of the cell is discarded in SCNT. An egg cell's nucleus is taken out at the same moment. The somatic cell's nucleus is subsequently transferred into the enucleated egg cell. The host cell reprograms the somatic cell nucleus after it has been injected into the egg. A shock stimulates the egg, which is now home to a somatic cell's nucleus, and causes it to start dividing. This one cell undergoes numerous mitotic divisions in culture before dividing into a blastocyst, an early stage embryo with roughly 100 cells that shares almost the original organism's entire DNA.

A somatic cell's nucleus (DNA) is transferred into an enucleated metaphase-II oocyte during somatic cell cloning (also known as cloning or nuclear transfer) in order to create a new individual who is genetically identical to the somatic cell donor. Science has undergone a revolution as a result of the success of creating Dolly, a whole animal, from a differentiated adult mammary epithelial cell. It proved that nuclear reprogramming—the process of returning a differentiated nucleus to a totipotent state—could entirely reactivate genes that had been inactivated during tissue differentiation.

Somatic cloning can be used to create numerous genetically superior farm animals, create transgenic animals for the manufacturing of pharmaceutical proteins or xenotransplantation, or it can be utilized to protect endangered species. It also holds tremendous medicinal promise for therapeutic cloning and allo-transplantation with optimization. The study of gene function, genomic imprinting, genomic reprogramming, regulation of development, genetic disorders, and gene therapy, among many other subjects, has made cloning an indispensable technique in addition to its practical applications. However, the low efficiency and high prevalence of developmental defects associated with cloning make it one of the most challenging problems to solve. The current nuclear transfer efficiency ranges from 0 to 10%, or 0 to 10 live births for every 100 transferred cloned embryos. Every research team investigating somatic cloning has come across developmental problems, including anomalies in cloned fetuses and placentas, in addition to high rates of pregnancy loss and neonatal death. It has been suggested that insufficient epigenetic signal reprogramming may be substantially responsible for low cloning efficiency.

Management of reproduction for optimum efficiency

Heat detection and mating

Cows are unable to engage in interactive sexual behaviour if they are always exhausted. Physical symptoms alone so meet the criteria, and identifying them take both motivation and knowledge. Cows should be inspected in such systems at least three times daily. Sometimes the vulva may enlarge and clear mucus may discharge as the only symptoms. Even the latter is only visible in buffalo when the animal is lying down. Confirmation can also be gained by watching the animal's reaction to pressure applied to the lumbar region or clitoral region. Observations for other estrous behaviours should be made at least three times per day, for a minimum of 15 to 20 minutes each time, in systems where there are many cows and they are permitted to contact during pasture or milking time. Regardless of whether they had been mated or not, cows that had earlier gone into heat should receive special attention. From the 18th day following the previous heat, such animals must be watched for a potential return to heat. If they begin to exhibit heat indications once more, it normally signifies that they have not given birth and must be mated once more. A record of when a cow has previously been in heat is the most crucial tool for heat detection.

Once heat has been discovered, the following step is to make sure the cow is mated at the ideal moment, which is between 12 and 18 hours after the commencement of heat or in the latter portion of the heat cycle. If a bull is used for hand-mating, two services separated by six to twelve hours are advised. When artificial intelligence (AI) is employed, cows that are found to be in heat in the morning should be served in the afternoon of the same day, and cows that are found to be in heat in the afternoon or evening should be served the following morning. For at least a week after mating, mated cows should not be subjected to excessive agitation, rough handling, or heat stress because it has been amply demonstrated that a rise in body temperature during this time might cause the death of the embryo. Repeat mating should be done if a cow continues to exhibit signs of heat for more than 12 to 24 hours following mating. The presence of blood discharge in a non-pregnant heifer or cow that has not been in heat for the previous three to four days (met-oestrus bleeding) typically indicates that the animal either experienced a silent ovulation or that the farmer failed to notice the animal's heat. These animals need to be monitored carefully once more beginning 17 days later.

Care of pregnant and parturient animals

Cows that are pregnant should be handled carefully to avoid stress and mishaps. They must be fed well during the dry time to facilitate the formation of body reserves, and they must be dried off two months prior to the anticipated calving date (eight months after conception for buffalo and seven months after conception for cattle). If at all feasible, wash the cows that will be giving birth and relocate them to a clean, separate enclosed area. Since reproductive issues that arise during pregnancy or close to parturition can occasionally be quite serious, immediate veterinarian intervention is recommended.

The postpartum period

After calving, vulval secretions should stop after around 15 days. Their continued presence suggests either a delayed involution or a uterine infection that needs to be treated. These cows are more likely to experience this if their calves were born abnormally. Within one month following calving, cows with excellent nutrition and management may start ovulating. However, it is advised to miss this heat and serve at the following one if the first heat occurs before 45 days have passed since calving. The cow can be served after 45 days, at the first heat, if the calving and uterine involution went off without any problems. The time between mating and calving should be roughly 90 days in cows that had postpartum infections or problematic calving.

If any of the following symptoms are present in heifers or cows, veterinarian treatment should be sought in order to preserve the best reproductive health in the herd:

- Abnormal discharges continuing for more than 15 days following calving;
- No heat being noticed 60 days following calving (90 days in some systems)
- No pregnancy after more than three additional services
- Anomalous discharges that occur during times of heat, such as flaky mucus, or at other times, such as pus
- Problems calving, placenta retention, or other issues with reproduction.

16

Milking Management

Letting the cow to release its milk so the dairy farmer can extract it for personal use or to sell is the procedure of milking. Consequently, it is not totally a natural process. The natural process must be manipulated by the dairy farmer to his advantage. Therefore, in order to control the natural process, one must first understand it. The process of producing milk from blood-derived basic materials by alveolar cells and storing it in the alveolar cavity is known as milk secretion. Only when the alveolar cavity is full and the pressure on the alveolar cells prevents further secretion does this continuous process come to an end. Milk can be withdrawn by the calf or the human milker by the process of "milk ejection," in which milk is released from the alveoli and runs into the ducts, the gland cistern, and the teat cistern. Hormones regulate the milk ejection reflex, also known as the "let down" reflex. The nerve endings in the teats provide signals to the brain when the calf suckles her mother, or when the milker washes the udder or begins to milk the cow. The posterior lobe of the pituitary gland, which is located in the back of the gland, gets activated when the message reaches the brain. Oxytocin, the "let down" hormone, is released into the circulation, travels to the udder, and induces the myoepithelial cells to contract that encircle the alveolus. The process of "letting down" will be interrupted by excitement, tension, pain, or fear. When a cow is in an uncomfortable circumstance, the adrenal cortex releases the hormone adrenaline. Oxytocin's effects are fully blocked by adrenaline. As a result, it is impossible to obtain significant milk outputs when the cow is milked in an environment where it is likely to be alarmed, scared, injured, or stimulated.

Frequency of milking

It is very popular that giving cows three daily milkings instead of two will result in significant improvements in milk production. Some of the data about three-times-per-day milking that has been gleaned through studies of milk recording data requires caution. In this instance, farmers have chosen which cows would receive three daily milkings. The highest-producing cows are chosen by farmers to receive three milkings daily, while the lower-producing cows receive only two. Milking four times daily results in a slight improvement over milking three times daily. There has been some interest in limiting the number of milkings with the goal of minimizing the labour required for milking.

Milking routine

Understanding the milk ejection process clarifies how the acts that make up the milking routine can train cows to let down their milk. Every milking should follow the same protocol in the milking shed to reinforce the conditioned "let down" reaction. As an illustration, bring the cow inside the shed, place it in the stall, feed it, clean the udder, and begin milking. The handling (cleaning) of the udder must be the final step in the process before milking begins. The sequencing of the remaining components of the routine is up to the farmer, but each milking should have a consistent flow. Some cows may have trouble conditioning the let-down response, especially if they have a significant amount of Bos indicus ancestry. The presence of the calf and the calf sucking naturally stimulate the let down response. It is typically not necessary for the cow's calf to be present when using highly selected dairy breeds of Bos taurus. It is occasionally necessary to have the cow's calf close before let down may be performed in Bos indicus breeds and Bos indicus crossbreeds, and with some individuals, it is necessary to let the calf nurse the cow for a brief period of time before the let down response takes place. The time it took to milk each cow is significantly decreased by just altering the procedure, and the amount of milk the cows produced increased as well. Each cow should be cleansed before being milked. Each cow should be fed before being scrubbed from head to udder as it entered the milking stall. The cows responded favourably to this particular milking schedule by producing outstanding yields.

Hygiene

Milking is also impacted by udder diseases and bacterial contamination of milk. Mastitis is the most typical illness in dairy cows. It most frequently occurs while milking. The number of cows that get mastitis is significantly reduced in areas free of dirt. Additionally, a milking practice that maintains udder cleanliness and creates a barrier of disinfection around the streak canal is crucial to prevent infection. The streak canal does not instantly close after milking, leaving the udder exposed to infection. This highlights the critical need of the disinfection barrier.

A sufficient supply of clean water is necessary for a clean environment. A system to collect and store the water that falls on the roofs of sheds and other structures can frequently be used to solve the issues of mud surrounding the yards and sheds as well as a supply of clean water. The first necessity for

17

Dairy Cooperative

Operation Flood and dairy cooperatives became India's greatest rural employment programme, enabling the modernization of the dairy industry to a point from where it may take off to fulfill the need for milk and milk products within the nation as well as chances on the international market. An unorganized sector handles 80% of the milk a rural producer produces, and an organized sector handles the remaining 20%. It has been discovered that dairy cooperatives significantly contribute to reducing rural poverty by increasing rural milk production and marketing. The main factors limiting milk producers' ability to negotiate better prices include the use of middlemen, a lack of infrastructural facilities for collecting, storage, transportation, and processing, and a lack of bargaining strength on their part. Future challenges for India's milk marketing include milk quality, product development, infrastructure support development, and international marketing.

One of the top producers of milk and dairy products worldwide is India. For a child's development and an adult's health, the nation's milk and dairy products are essential. Milk is the ideal food for babies, it is crucial for developing children, it is the most vital food for adults, and it is the most important food for expectant mothers. In India, it is typically noticed that cows are better milkers in the north and west, but that the quantity of the milk produced declines as one travels toward the south and east. In a nation like India, the significance of the animal industry in general and dairy farming in particular barely needs to be emphasized. Production of milk and milk products is the livestock industry's primary and best-known contribution. This industry also contributes hides, skins, organic manure, and draught power in addition to milk. Different-sized independent farms are involved in the milk production process. A better system for assembling, processing, and distributing milk and
s products is severely constrained by this aspect of the milk production system
the perishable nature of milk. The co-operative movement in India must
into consideration for the research of dairy activities to be thorough.
team to accomplish goals is what is meant by cooperation. The
village communities was predominately focused on the idea
ntrast is established between them based on the amount

of territory they cover at the local, regional and national. Dairy production is promoted cooperatively in many Indian states. The strong backing of farmers and the great attention of the state administration are the key drivers of cooperative dairy development in the state. The primary village cooperative societies are connected to the district union and state federation, which are governed by the national co-operative dairy federation in India, as part of the three-tier framework that today forms the foundation of the dairy industry.

Everywhere at the district level, dairy activity is done. The main milk organizations gather milk at the village level and provide it to the dudh sangh at the district level. This includes the district's milk collection and dudh sangh distribution.

- To integrate the disorganized dairy industry into the organized one.
- To enhance animal health care and input activities to boost milk production
- To promote the availability of liquid milk.
- To maximize the state's milk production potential and to guarantee steady growth in marketing and procurement.
- To guarantee that milk and milk products meet the legal requirements for quality.
- The major oversees and controls dairy-related activity.

In the previous three decades, India's annual milk production has more than tripled. The Operation Flood (OF) Project's support from dairy cooperatives is significantly attributed for this country's quick development and modernization. Government and cooperatives are part of the organized sector; private businesses are part of the unorganized sector. Dairy cooperatives stand out from other milk marketing channels due to their involvement in the acquisition of milk and the provision of essential services to dairy producers. Dairy farmers that sell their milk to dairy cooperatives are paid fairly for it. Additionally, these facilities offer monetary security and periodically distribute funds to dairy farmers. As a result, the dairy cooperatives give the dairy producers a combined sum. The primary barrier to this channel is the dairy cooperatives' relatively slow payment schedules. The impoverished households prefer trade their marketable surplus through other channels since they cann longer times to receive the cash.

It is clear that dairy farming is crucial to the livelihood of i landless people in the local economy. Government and co are involved in milk marketing through the organized sector, whilst private groups are involved in the unorganized sector. The unorganized sector accounts for the majority of milk marketing, with the organized sector having a very small contribution. The "operation flood" programme, launched by the Indian government, aims to properly count milk and milk derivatives. Various initiatives were also made to encourage the creation of dairy cooperatives. India's dairy cooperatives are organized in three tiers, with state-level federations, district milk unions, and village-level dairy co-operative societies, respectively. Dairy co-operative societies have been established as a result to supply the local milk demand and end the current milk marketing fraud. The establishment of dairy cooperatives and the construction of their infrastructure cost a substantial sum of money. The growth of dairy cooperatives has fueled rural people's empowerment and aided rural development. In the future, dairy cooperatives should contribute greatly to revitalizing rural life. In the era of liberalized environments, the cooperative sector needs to act quickly to take advantage of every opportunity that presents itself. Low capacity utilization of dairy plants, non-viability owing to losses, a lack of operating capital, and an inappropriate marketing strategy to compete with the market due to inadequate training and knowledge are the main issues cooperatives confront. In order to strengthen cooperative operations, increase productivity, ensure quality, expand dairy infrastructure, and create a national information network by implementing various programmes, the government should offer technical and financial support to dairy cooperatives.

Supporting Dairy Cooperatives and Farmer Producer Organizations engaged in dairy activities:- A Scheme named "Supporting Dairy Cooperatives and Farmer Producer Organizations engaged in dairy activities" was approved to provide working capital loan to State Cooperatives and Federations (Source and for detail: https://dahd.nic.in/schemes/programmes/sdcfpo)

Objectives

- To assist the State Dairy Cooperative Federations by providing soft working capital loan to tide over the crisis on account severely adverse market conditions or natural calamities.
- To provide stable market access to the dairy farmers.
- To enable State Cooperative Dairy Federations to continue to make timely payments of dues to the farmers.
- To enable the cooperatives to procure milk at a remunerative price from the farmers, even during the flush season.

On account of economic difficulties faced by the dairy cooperatives and Producer Owned Institutions due to Covid-19 lockdown, it was decided to incorporate the component of "Interest Subvention on working capital loans" under the above scheme. The products covered under the scheme for availing working capital loans are Skimmed Milk Powder (SMP), Whole Milk Powder (WMP), White Butter and Ghee. Supporting Dairy Cooperatives and Farmer Producer Organizations engaged in dairy activities (SDCFPO) as a part of Umbrella Scheme "Infrastructure Development Fund" from 2021-22 to 2025-26.

In the years to come, the dairy cooperatives must have a significant impact on the economy of our country. The milk industry is expected to experience a new boom. The industry's significant role in creating newer, direct and indirect employment opportunities as well as its impact on raising the nutritional standards of our population both increase the significance that should be given to this sector in the twenty-first century. The cooperative can reach new heights in the near future with the right encouragement and member support.

18

Carbon Foot Print of Milk

The cradle-to-grave dairy supply chain includes a number of operations that result in greenhouse gas (GHG) emissions related to milk production. These activities broadly include the use of agricultural inputs like fertilizers, fossil fuel for the production of animal feed and fodder, processing concentrates in cattle feed plants, enteric fermentation, manure management, and post farm gate activities like processing milk at dairy cooperative societies, milk unions, its conversion into different milk products, and their distribution through retail supply chains. As a result, it's crucial to take into account every step of the dairy supply chain and calculate the milk's "Carbon footprint" using the widely used Life Cycle Assessment (LCA) approach. This offers a possibility to identify GHG emission hotspots and develop mitigating strategies for the dairy industry. The total net GHG emissions during the course of milk's lifecycle inside a particular system boundary and in relation to a specified amount of milk with a specified composition make up the carbon footprint (CF) of milk. Enteric fermentation is the main hotspot, accounting for up to 75% of the dairy industry's overall GHG emissions. The Ration Balancing method used by NDDB has shown promise in lowering enteric methane emission from cows and buffaloes. Given the long-term advantages of ration balancing on milk output, reproduction, and general animal performance, it is predicted that providing balanced rations can incresse cow and buffalo milk production by up to 30%.

Rice cultivation, enteric fermentation, and manure management together account for 61% of India's total methane emissions, with enteric fermentation contributing 40%, rice cultivation contributing 17%, and manure management contributing 4%. Methane is produced as a byproduct of the feed digestion system in the rumen, a sizable forestomach, where the fermentation process takes place. Methanogenesis, also known as biomethanation, is a microbial metabolic process of anaerobic respiration in which methane is created with the help of the methanogen microbes such as bacteria, viruses, fungi, protozoa, and others by reducing carbon dioxide by hydrogen. Since the majority of these studies are theoretical calculations and devoid of any systematic experiments, their estimates of methane emission show wide variability in the output by Indian cattle. Sheep and goats generate roughly 30 to 40 litres per day, whereas cattle and buffalo typically produce 200 to 250 litres. Cattle and buffalo account for the majority (10.9 Tg/yr) of the livestock's 11.75 Tg/yr total methane output, whereas other livestock only contribute 0.86 Tg/yr. Along with the response from manure piles, which contributes far less methane than enteric fermentation, enteric fermentation also contributes the majority of methane to the atmosphere. In the case of manures, the bacteria involved in the methanogenesis process break down the cellulose content of the manure. Before being used to agricultural land, manure produced by animal production systems is often dumped as such at one location in solid form. Low dietary nutrient digestibility may make the dung more likely to produce more fermentable organic matter, which can increase the quantity of methane the manure produces. In some cases, the wastes are also composted before being spread on farmland or anaerobically digested to create methane gas for use as a biofuel for cooking. Livestock is a significant source of methane emissions in the agriculture sector in India. Methane emissions from manure management and enteric fermentation make for 9% and 91% of the total. Cattle and buffalo are the greatest emitters of methane when compared to other species. Methane emissions from cattle are a substantial source of anthropogenic greenhouse gas emissions in many nations, and efforts are being made to find ways to cut these emissions. Several methods are currently being investigated and developed. Each technique has a unique cost, cost-effectiveness, and end-user adoption profile. Despite the fact that ruminant food and nutrition can significantly affect methane production. It is both practical and expensive to alter these animals' diets to lower their absolute methane emissions. On an experimental level, initiatives to reduce ruminant methane synthesis using antibiotics, biotechnological methods like rumen defaunation, methanogenesis inhibitors like PSM. Additionally, in vivo animal experiments should be used to test such efforts to lower methane production in order to ascertain the best doses for inhibiting methane emission while having the least detrimental effect on ruminal fermentation of feed, so that they can be used in practice for affordable and environmentally friendly livestock production. Already, ruminant enterprises are using traits like live weight increase, milk output, feed efficiency, and genetic variety to indirectly reduce their emissions intensity.

19

Future Sciences for Indian Dairy

Electronic nose for estrus detection

The Indian economy, which meets its own dairy demands, has reached a turning point with its bovine breeding initiatives. The viability of this enterprise, however, depends largely on completing as many successful artificial insemination cycles at a cattle farm. Estrus behaviour observation is used to determine when artificial insemination should take place. But since not all cattle exhibit substantial estrus symptoms, insemination results are not always reliable. An estrus detecting instrument known as an Electronic Nose (eNose) is created to address these issues. Electronic nose (E-Nose) technology has been used as a substitute method for detecting changes in perineal odour linked to estrus in cows. This is mostly based on observations of estrus cycles, which present cow with numerous chances to conceive throughout their productive lifetimes. Since inadequate estrus detection is a significant contributor to cow infertility and incurs significant financial loss to the dairy industry, the proportion of cows that become pregnant during a breeding season directly affects the profitability of cattle farms. Although the herdsperson's behavioural observation of estrus is a major factor in determining the timing of artificial insemination, success rates are only about 50%. Numerous research have discussed potential techniques for enhancing estrus detection, including as measuring vaginal impedance, using trained teaser bulls, measuring milk and body temperatures, and monitoring activity. None of these techniques, though, has produced data that is as simple and reliable to reproduce. Observing estrus behaviour or using the findings of a milk progesterone assay, both of which require a lot of labour and have low fertility rates, are still the main methods used to identify estrus.

Sexed semen

Sexed semen is described as having X or Y bearing sperm to generate offspring of a desired sex, either female or male (with roughly 90% accuracy). Researchers with the USDA in California created the technique for gender classification. Sperm are sorted by recognizing distinctions between sperm that bears X and Y chromosome. In cattle, the X-chromosome (female) has roughly 3.8% more DNA than the Y-chromosome (male). The X- and Y-bearing sperm are separated using the differential in DNA content. The most effective method for semen sexing has been found to be flow cytometry-based sorting. Over the years, the technology has been improved to the point that sex sorting can now be done with a purity of more than 90%. The method has been patented, standardized, and commercialized in the USA, Europe, and other nations. Sexed semen can be used for artificial insemination without risk. The conception rate with sexed semen is reportedly 10 to 15% lower than with conventional semen because the concentration of sperm in the sexed semen straw is significantly lower than that in the conventional semen straw and the sorting technique itself harms the sexed sperm.

Advantages of using sexed semen

The producers are able to conserve resources by raising just female calves rather than undesirable males. The availability of replacement heifers grows as more female calves are produced. There is a chance to sell excess heifers to different farmers or farms. Through improved IVF, embryo transfer, and progeny testing programmes, this technology may hasten genetic advancement. By acquiring heifers from outside sources, which increases biosecurity, this is an economical way to increase herd size without running the danger of spreading diseases. Only viable sperm are available because during the sorting process, dead, dying, or damaged sperm cells are eliminated, allowing the sexed semen to be successful even at a lower concentration than traditional semen. It will be easier to give birth to female calves than male calves by employing more sexed semen to produce more female calves. This is especially helpful for young heifers.

Limitations of using sexed semen

Sex-sorted semen has a 10-15% lower conception rate than regular semen. Given the limited coverage (20–25%) and low conception rate (25–35%) of artificial insemination in India, this element will be even more important. In comparison to conventional semen, which contains 20 million sperm each dosage, sexed semen has between 2 and 4 million sperm per dose. Under Indian conditions, managing reduced sperm concentration will be difficult. The price of the sexed semen is depending on the breed and pedigree of the bull used to produce the sexed semen as well as the quantity purchased. Sexed semen should be utilized in heifers due to the high fertility rate of heifers, which will increase the likelihood of pregnancy. However, it can also be utilized in cows up to their third lactation that have a strong track record of reproduction.

Artificial insemination in goat

Artificial insemination (AI) is a straightforward procedure that has found success in the breeding of dairy cattle and buffalo. Its assistance in enabling India to produce the most milk in the world cannot be disregarded. In case of goat farming, due to early castration and the killing of male goats for meat, there are fewer good breeding bucks available. It is economically unviable to maintain a breeding buck for a small flock because goats are typically raised by small, marginal, and landless farmers, with flock sizes of 3 to 5 goats. As a result, unimpressive males breed with the females, causing the region's valuable genetic diversity to be lost. The Eastern Regional Station of the ICAR-National Dairy Research Institute, Kalyani, West Bengal, has worked on the preservation of Bengal goat semen, standardization of artificial insemination technique, and transfer to the field in order to prevent the dilution of valuable germplasm by indiscriminate mating, preserve its purity, and propagate this unique germplasm. For the preservation of goat sperm, a soybean-lecithin-based animal-source free semen extender was created. With a 47.26% conception rate, liquid semen has a storage life of roughly 72 hours when kept at a cold temperature. While in the field, frozen semen straws have a conception rate of about 46.52%. Presently Mathura Veterinary University is doing good work on artificial insemination technology in goat.

Non-Bovine Milk from Animal Sources

At 96% of the overall production, cow and buffalo milk still make up the majority of the milk. In 2016 and 2020, cow and buffalo milk, contribute 82.6% & 13.9% and 81% & 15.2% respectively of the world's milk production. In 2020, the proportion of goat, sheep, and camel milk production in the world was 2.3%, 1.2%, and 0.4%, respectively. Even though they are mostly utilized in processed dairy products and are essential to the rural economy, together, these later species produced less than 5% of the milk in the world. In some tribes, mare, donkey, and deer milk are very important culturally even though they are not frequently consumed or traded. Non-bovine milk has recently been the focus of significant research projects due to its possible health benefits. The composition of milk is influenced by a variety of animal characteristics (species, age, parity, lactation stage, breed, and overall genetic background and health), agricultural practices (management system, nutrition, and the number of offspring), and environmental factors. To meet the demands of their offspring, various species' milks have evolved in composition and physicochemical properties. The physicochemical characteristics of milk, which will affect its functional properties and a number of consequent technological and sensory aspects, as well as the value of the milk and its products, are taken into consideration when milk is utilized for human consumption. The impact of sheep breeds on the calibre of dairy products is not well understood.

According to a recent study, animal milk has probably been consumed by humans as a food source for a very long period, making milk in general a food with considerable global economic value. Because animal milk contains high levels of pure liquid, sugar, fat, B vitamins, and calcium, it is an excellent source of nutrition. Humans have adapted behaviorally and physiologically to the drinking of milk through the creation of techniques like pasteurization and curdling as well as biologically through the emergence of lactase persistence that facilitates the digestion of lactose in milk.

Nutritional composition of non-bovine milk

Dairy goats are still largely found in low-income, food-insecure areas of India, where their products represent an important food source, despite the fact that the demand for non-bovine milk is now growing. The reasons for the increase in demand include dairy allergy and intolerance detection as well as escalating consumer desire for "healthier" meals. Goat's milk is an alternative for those who are sensitive to or allergic to cow's milk. An abnormal reaction to the proteins in cow's milk characterizes cow's milk allergy, a common paediatric condition. Since between 25 and 40% of persons who are allergic to cow's milk can consume goat's milk, goat's milk may occasionally be used as a substitute. Although all ruminant milk proteins share a similar structure, not all people with a cow's milk allergy experience this. Although all milk comes from ruminants, there are significant differences in the nutritional composition of cow, goat, and sheep's milk.

It is vital to keep in mind that, in addition to the age of the animal and its stage of lactation, nutrition and breeding can have a considerable impact on the composition of ruminant milk. In general, goat's milk has more short- and medium-chain fatty acids, which are good for creating cheese, and larger proteins and smaller fat globules than cow's milk. Sheep's milk has good yields for manufacturing cheese because it is rich in proteins, minerals (calcium, phosphate, and magnesium), and lipids (mainly medium chain). When compared to cow's milk, goat's and sheep's milk has incredibly low quantities of agglutinin.

Sheep's milk has a higher protein and fat content than cow's and goat's milk, as well as slightly more PUFA and CLAs. In addition, compared to cow's milk, it contains high levels of calcium (important for bone formation and health), phosphorous (for healthy bones and teeth), magnesium (for a healthy immune